Bifunctional Compounds

Robert S. Ward

University of Swansea

OXFORD   NEW YORK   TOKYO
OXFORD UNIVERSITY PRESS
1994

Oxford University Press, Walton Street, Oxford OX2 6DP

Oxford   New York   Toronto
Delhi   Bombay   Calcutta   Madras   Karachi
Kuala Lumpur   Singapore   Hong Kong   Tokyo
Nairobi   Dar es Salaam   Cape Town
Melbourne   Auckland   Madrid

and associated companies in
Berlin   Ibadan

Oxford is a trade mark of Oxford University Press

Published in the United States
by Oxford University Press Inc., New York

A catalogue record for this book is available from the British Library

Library of Congress Cataloging in Publication Data
Ward, Robert S.
Bifunctional compounds / Robert S. Ward.
(Oxford chemistry primers; 17)
Includes bibliographical references and index.
1. Organic compounds.   I. Title.   II. Series
QD255.4.W37 1993   547–dc20   93-33958

ISBN   0-19-855809-0 (Hbk)
ISBN   0-19-855808-2 (Pbk)

Typeset by the author and AMA Graphics Ltd., Preston, Lancs
Printed in Great Britain by
The Bath Press, Avon

Series Editor's Foreword

Bifunctional compounds are ideal substrates for the understanding of reactivity and hence synthesis. Their chemistry exemplifies – and thus allows a detailed understanding of – the fundamental interactions of the various functional groups.

Oxford Chemistry Primers have been designed to provide concise introductions relevant to all students of chemistry, and contain only the essential material that would usually be covered in an 8–10 lecture course. This tenth organic primer not only provides a basic understanding of functional group reactivity, but also provides a good basis for revision. This primer will be of interest to apprentice and master chemist alike.

Stephen G. Davies
The Dyson Perrins Laboratory, University of Oxford

Preface

Bifunctional compounds occupy a central position in organic chemistry. They are important as starting materials for organic synthesis and as templates for studying organic reaction mechanisms. They are, for example, used as precursors for the preparation of carbocyclic and heterocyclic compounds. Furthermore, the products of synthesis and the products of nature are usually polyfunctional compounds and an understanding of the interaction between functional groups is vital to any attempt to explain or control their behaviour.

This book outlines some of the methods used to prepare bifunctional compounds and goes on to describe the chemistry of some of the more important classes. The section dealing with the preparation of bifunctional compounds provides a useful revision of the reactions of monofunctional compounds. This is in keeping with the position occupied by bifunctional compounds in most chemistry courses, namely bridging the gap between an elementary treatment of monofunctional compounds and the more advanced aspects of the subject.

The sections dealing with the chemistry of enamines, enol ethers and enolates give just a flavour of the modern methodology of organic synthesis. The chapter dealing with protecting groups is particularly relevant to the manipulation of carbohydrates and amino acids. Another important use for bifunctional compounds is as precursors for the preparation of polymers. Due to limitations of space this aspect is only briefly mentioned.

Finally, I should like to express my thanks to Dr Mike Williams and Dr Stephen Davies for reading and commenting on the first draft and for several helpful suggestions.

Swansea   R.S.W
June 1993

Contents

1 Introduction

A large number of important organic compounds contain two or more functional groups. In many cases the chemistry of such compounds is similar to that of the corresponding monofunctional compounds, but in other cases the presence of two functional groups, particularly when they are in close proximity, gives rise to modified or even unique chemical and physical properties. For example, the carbon–carbon double bond of an αβ-unsaturated carbonyl compound readily undergoes addition by nucleophiles (eqn 1.1), in contrast to the more usual situation in which alkenes react only with electrophiles. Conversely, the carbon–carbon double bond of an enamine is so strongly nucleophilic that it will even react with alkyl halides (eqn 1.2).

(1.1)

(1.2)

Thus, in some cases the chemical properties of one functional group are dramatically altered by the presence of a second functional group. In other cases, the presence of a second functional group may significantly alter the rate of a reaction (eqns 1.3 and 1.4).

$$CH_3CH_2CH_2Br + H_2O \xrightarrow{\text{slow}} CH_3CH_2CH_2OH + HBr \quad (1.3)$$

$$CH_2{=}CHCH_2Br + H_2O \xrightarrow{\text{fast}} CH_2{=}CHCH_2OH + HBr \quad (1.4)$$

Allyl bromide reacts rapidly by an S_N1 mechanism since a stable carbocation is produced, see section 9.3, p.70.

Even when two functional groups are widely separated they may still interact with each other. When they do so intramolecularly it leads to cyclic products. For example, diketones and diesters undergo intramolecular aldol and Claisen condensations under basic conditions leading to cyclic products (e.g. eqns 1.5 and 1.6). When functional groups interact intermolecularly it leads to polymeric products (e.g. eqns 1.7 and 1.8).

This book is concerned with the chemistry of compounds in which the presence and interaction of two functional groups gives rise to properties which are different from those of the corresponding monofunctional compounds.

Aldol condensation

$$\text{(1.5)}$$

Claisen condensation

$$\text{(1.6)}$$

$$HO(CH_2)_{10}CO_2H \xrightarrow{H^+} \quad \text{polyester} \quad \text{(1.7)}$$

$$\xrightarrow{\Delta} \quad \text{polyamide} \quad \text{(1.8)}$$

1.1 Classification and nomenclature

Bifunctional compounds can be classified according to the nature of the groups present and their relative position. Some of the main types are shown in scheme 1.1. In addition, it is relevant to include certain vinyl and allyl derivatives, and heterocumulenes, which are shown in scheme 1.2. Indeed, enols, enol ethers, enamines, and vinyl halides can be regarded as analogues of the corresponding carboxylic acid derivatives. However, it is instructive to consider them as a subset of bifunctional compounds since they provide a supreme illustration of the way in which the chemistry of bifunctional compounds differs from that of their component parts. Similarly, allyl alcohols, ethers, amines, and halides can be compared with the appropriate α-substituted carbonyl compounds (see scheme 1.1). Finally, the heterocumulenes, such as ketenes, isocyanates, and carbodiimides, form a small, self-contained group of bifunctional compounds.

When a compound contains two (or three) identical functional groups it is named in the same way as the corresponding monofunctional compound except that 'di-' or 'tri-' is added to the appropriate prefix or suffix to indicate the presence of two (or three) such groups.

2,4-pentanedione

1,2-dibromoethane

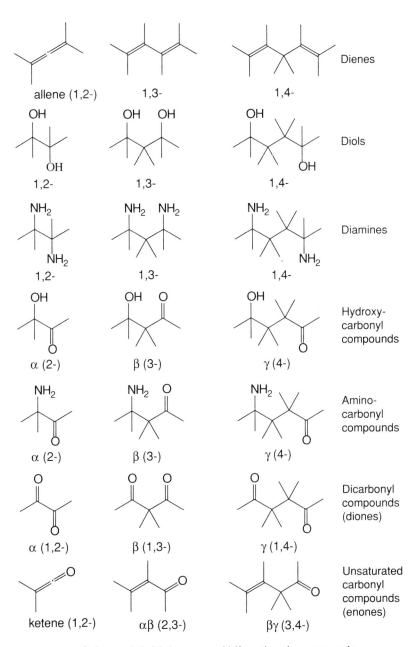

Scheme 1.1 Main types of bifunctional compounds

X = OH   enol   X = OH   allyl alcohol
X = OR   enol ether   X = OR   allyl ether
X = NR$_2$ enamine   X = NR$_2$ allyl amine
X = halogen   vinyl halide   X = halogen   allyl halide

Heterocumulenes

ketene   isocyanate   carbodiimide

Scheme 1.2

When a compound contains two (or more) different groups, usually only one of them can be indicated by a suffix and the others must be indicated by a prefix. The group which is indicated by a suffix is the one which has the highest priority as listed in Table 1.1. The alternative suffix shown in the table is used when naming derivatives of cycloalkanes and arenes.

methyl 4-oxopentanoate   5-amino-2-pentanol
not 4-methoxycarbonyl-2-butanone   *not* 4-hydroxy-1-pentanamine

Table 1.1 Names of functional groups in order of decreasing priority

Group	Prefix	Suffix	Alternative suffix
$-CO_2H$	carboxy-	-oic acid	-carboxylic acid
$-SO_3H$	sulpho-	-esulphonic acid	
$-CO_2R$	alkoxycarbonyl-	-oate	-carboxylate
$-SO_3R$	alkoxysulphonyl-	-esulphonate	
$-COCl$	chloroformyl-	-oyl chloride	-carbonylchloride
$-CONH_2$	carbamoyl-	-amide	-carboxamide
$-CN$	cyano-	-enitrile	-carbonitrile
$-CHO$	oxo- (or formyl-)	-al	-carbaldehyde
$-C=O$	oxo-	-one	
$-OH$	hydroxy-	-ol	
$-SH$	mercapto-	-ethiol	
$-NH_2$	amino-	-amine	
$-OR$	alkoxy-	—	
$-SR$	alkylthio-	—	
$-Cl$	chloro-	—	
$-NO_2$	nitro-	—	

As well as locating substituents by numbers, Greek letters are also sometimes used. This is a particularly common and useful way of referring to different classes of bifunctional compounds (see scheme 1.1), but it is not usually used when naming an individual compound. The Greek letter is used to indicate the separation between the two functional groups. Using this approach the first carbon atom away from the main functional group is denoted as the α carbon atom, the next is the β carbon atom, and so on.

α-amino acid   β-keto ester   β-diketone
2-aminohexanoic acid   ethyl 3-oxobutanoate   1,3-cyclohexanedione

2 Preparation of bifunctional compounds

2.1 General strategy

The methods used to prepare bifunctional compounds are usually specific to the particular combination of functional groups present. Nevertheless, it is useful to consider in general terms the strategy which must be adopted to prepare such compounds. For example, one way to approach the synthesis of many 1,2-bifunctional compounds would be to consider as a possible precursor the alkene which would result from eliminating the two functional groups (scheme 2.1). Many 1,2-bifunctional compounds can indeed be prepared from alkenes.

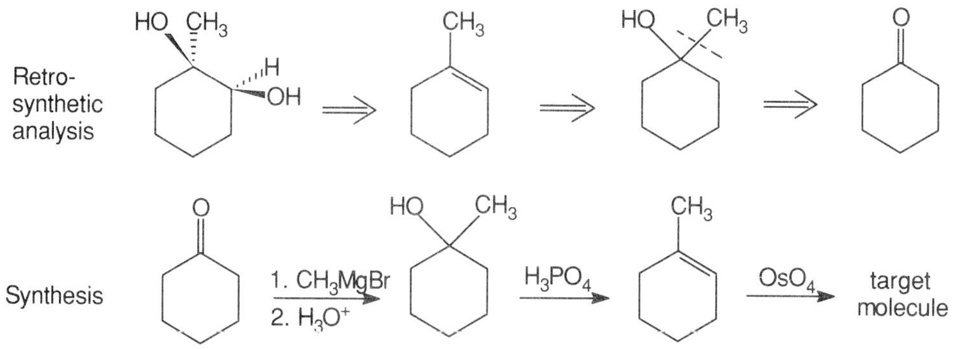

Scheme 2.1

This approach to organic synthesis is called *retro-synthetic analysis.* Scheme 2.2 shows some other *disconnections* which result from a retro-synthetic analysis of 1,2-bifunctional compounds. Thus, an alternative approach would be to imagine cleaving the bond linking the two functionalised carbon atoms, and in the case of a 1,2-diol, this would suggest a method involving the reductive coupling of two carbonyl components. As will be seen in section 2.2 this is a practical way of making many 1,2-diols.

Working backwards is a logical way to tackle the synthesis of a given target molecule. The value of this approach is that it leads one to select appropriate methods for bringing about the forward transformation. The individual reactions which utilise the disconnections shown in scheme 2.2 are considered in more detail in section 2.2.

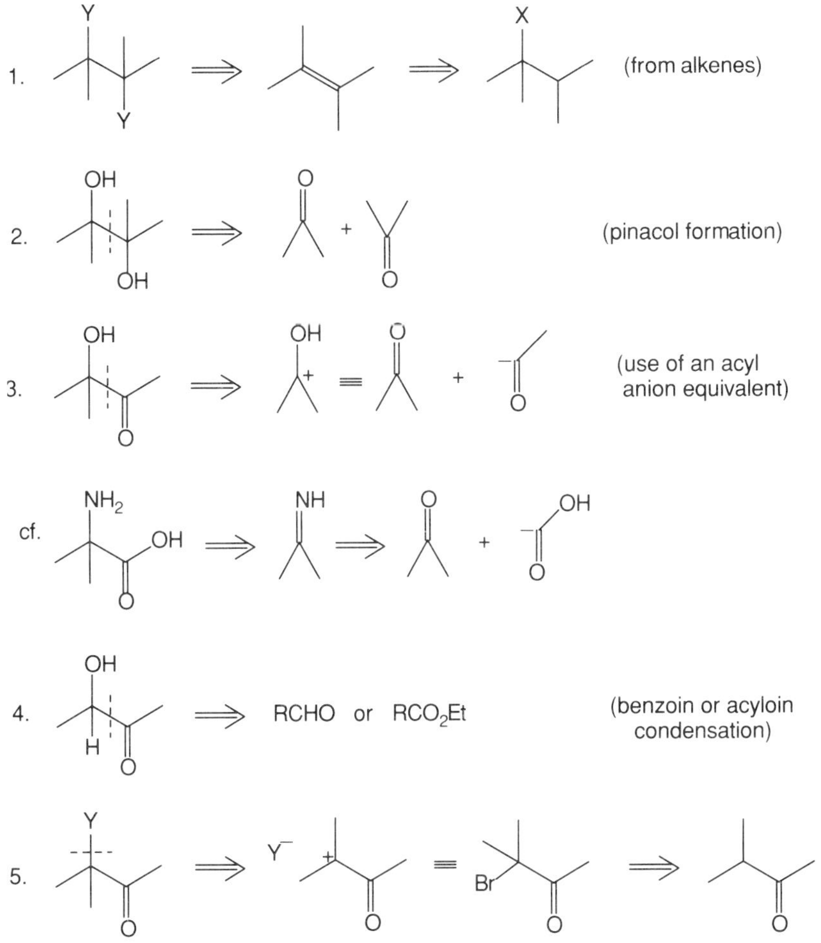

Scheme 2.2 Retro-synthetic analysis of 1,2-bifunctional compounds

2.2 Preparation of 1,2-bifunctional compounds

In summary, 1,2-diols may be prepared by hydroxylation of alkenes, by reductive dimerisation of ketones, or by nucleophilic addition to α-hydroxycarbonyl compounds. α-Hydroxycarbonyl compounds can in turn be prepared using an acyl anion equivalent, or by means of the acyloin or benzoin condensations. α-Hydroxyacids can also be prepared by way of the benzilic acid rearrangement (see section 6.1). α-Amino acids can be prepared from α-aminonitriles or from α-halocarbonyl compounds. They can also be prepared from diethyl malonate (see section 6.2). 1,2-Dicarbonyl compounds are readily prepared by mild oxidation of α-hydroxycarbonyl compounds.

From alkenes

Alkenes afford direct access to many 1,2-bifunctional compounds. Thus 1,2-diols can be stereoselectively prepared from alkenes, and epoxides offer many possibilities for the synthesis of other 1,2-bifunctional compounds (scheme 2.3).

Mechanism of epoxide formation

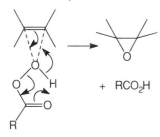

Mechanisms of epoxide opening

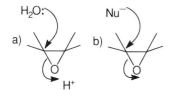

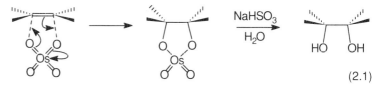

Scheme 2.3

The mechanism of the osmium tetroxide reaction involves the formation of a cyclic osmate ester which is usually hydrolysed, without isolation, to give the 1,2-diol. This reaction therefore results in *syn* addition of the two hydroxyl groups since the hydrolysis step does not involve any change in the configuration of either of the two carbon atoms (eqn 2.1). In contrast, epoxidation, followed by acid or alkaline hydrolysis, or nucleophilic addition, results in overall *anti* addition since the opening of the epoxide takes place with inversion of configuration at the carbon atom undergoing attack. The development of highly stereoselective methods for the synthesis of certain chiral epoxides (see section 9.2) has made possible the asymmetric synthesis of a wide variety of 1,2-bifunctional compounds.

$$(2.1)$$

Pinacol formation

Symmetrically substituted 1,2-diols can be prepared by reductive dimerisation of aldehydes and ketones (eqn 2.2).

$$(2.2)$$

The mechanism of this reaction involves the coupling of two radical anions formed by interaction of the ketone with a metal (eqn 2.3). A number of alternative reagents can be used (e.g. Na–Hg or SmI_2) and the reaction can also be brought about photochemically in the presence of an alcohol which acts as a source of hydrogen atoms.

$$2 \quad \text{(structure)} \xrightarrow{\text{Mg}} \text{(structure)} \longrightarrow \text{(structure)} \xrightarrow{H_3O^+} \text{product} \quad (2.3)$$

Use of an acyl anion equivalent

One way of preparing α-hydroxycarbonyl compounds is by using an *acyl anion equivalent*. This is a species which reacts as a carbanion but which can be subsequently converted into a carbonyl group. Some examples are shown in scheme 2.4.

Acyl anion equivalents effectively reverse the normal polarity of the carbonyl group they represent. This principle, of reversing the inherent reactivity of a functional group (e.g. bringing about reaction with an electrophile at a carbonyl carbon atom) is known as *umpolung*. There are many examples of the use of this principle in organic chemistry and indeed many other uses for acyl anion equivalents. Just as α-hydroxyacids can be prepared from cyanohydrins (scheme 2.4), so α-amino acids can be prepared from α-aminonitriles, which can be prepared from aldehydes and ketones (eqns 2.4 and 2.5).

acyl anion

$$\text{RCHO} \xrightarrow[H^+]{\text{NaCN}} \underset{H \quad CN}{\overset{R \quad OH}{\diagup\!\!\diagup}} \xrightarrow{H_3O^+} \underset{H \quad CO_2H}{\overset{R \quad OH}{\diagup\!\!\diagup}} \quad (2.4)$$

$$\text{RCHO} \xrightarrow[KCN]{\text{NH}_4Cl} \underset{H \quad CN}{\overset{R \quad NH_2}{\diagup\!\!\diagup}} \xrightarrow{H_3O^+} \underset{H \quad CO_2H}{\overset{R \quad NH_2}{\diagup\!\!\diagup}} \quad (2.5)$$

First step involves imine formation

$$\text{RCHO} \xrightarrow{NH_3} \text{RCH=NH} + H_2O$$

Acyl anion equivalents such as CN⁻ can also be used to prepare a wide range of dicarbonyl compounds (e.g. eqns 2.6–2.8).

$$\underset{CH_3 \quad Cl}{\overset{O}{\diagdown}} \xrightarrow{\text{NaCN}} \underset{CH_3 \quad CN}{\overset{O}{\diagdown}} \xrightarrow{\text{conc.HCl}} \underset{CH_3 \quad CO_2H}{\overset{O}{\diagdown}} \quad (2.6)$$

$$Cl\!\!\diagup\!\!CO_2H \xrightarrow{\text{NaCN}} NC\!\!\diagup\!\!CO_2H \xrightarrow{\text{NaOH}} HO_2C\!\!\diagup\!\!CO_2H \quad (2.7)$$

$$CH_3\!\!\diagup\!\!CO_2Et \xrightarrow{\text{NaCN}} \underset{CH_3}{\overset{CN}{\diagup\!\!\diagup}}CO_2Et \xrightarrow[H_2O]{^-OH} \underset{CH_3}{\overset{CO_2H}{\diagup\!\!\diagup}}CO_2H \quad (2.8)$$

Mechanism of hydration of alkynes

$$HC\equiv C^- \quad\equiv\quad \overset{\displaystyle }{\underset{\displaystyle O}{\Big\Vert}}\text{(acetone anion)}^-$$

e.g. $HC\equiv C^- \xrightarrow{Me_2CO}$ (tertiary alcohol with alkyne) $\xrightarrow[Hg^{2+}]{H_3O^+}$ (tertiary alcohol with ketone)

Mechanism of hydration of alkynes diagram

$$R\bar{C}HNO_2 \quad\equiv\quad \overset{R}{\underset{O}{\diagdown}}^-$$

e.g. $R\bar{C}HNO_2 \xrightarrow{Me_2CO}$ (nitro alcohol) $\xrightarrow[H_2O]{TiCl_3}$ (hydroxy ketone)

$$^-CN \quad\equiv\quad ^-CO_2H$$

e.g. NaCN $\xrightarrow[H^+]{Me_2CO}$ (cyanohydrin) $\xrightarrow[25°C]{conc.HCl}$ (hydroxy acid)

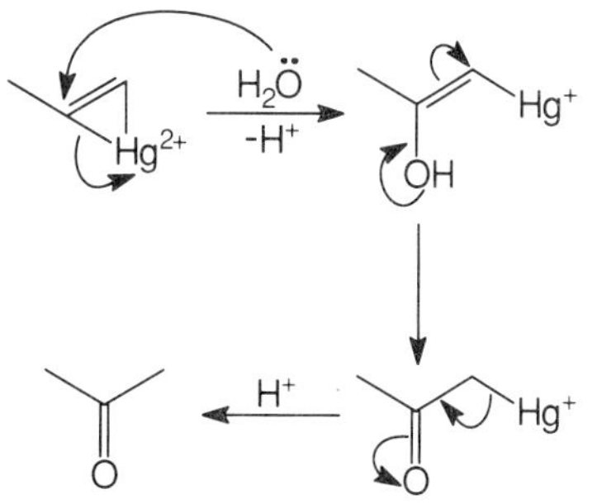

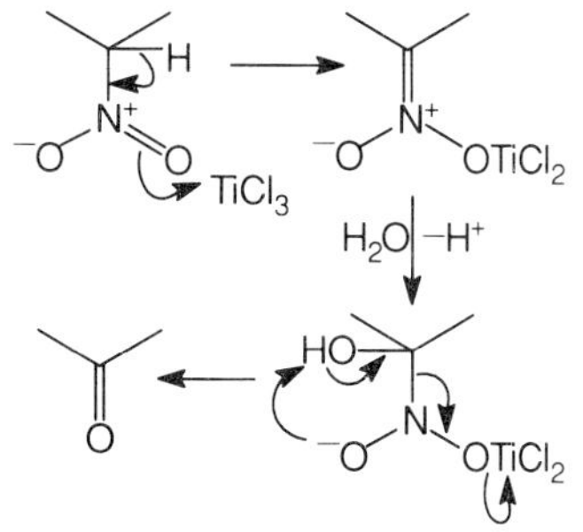

Mechanism of hydrolysis of dithiane

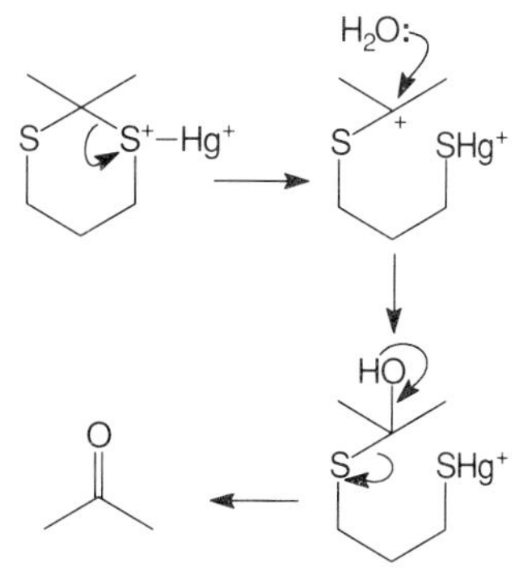

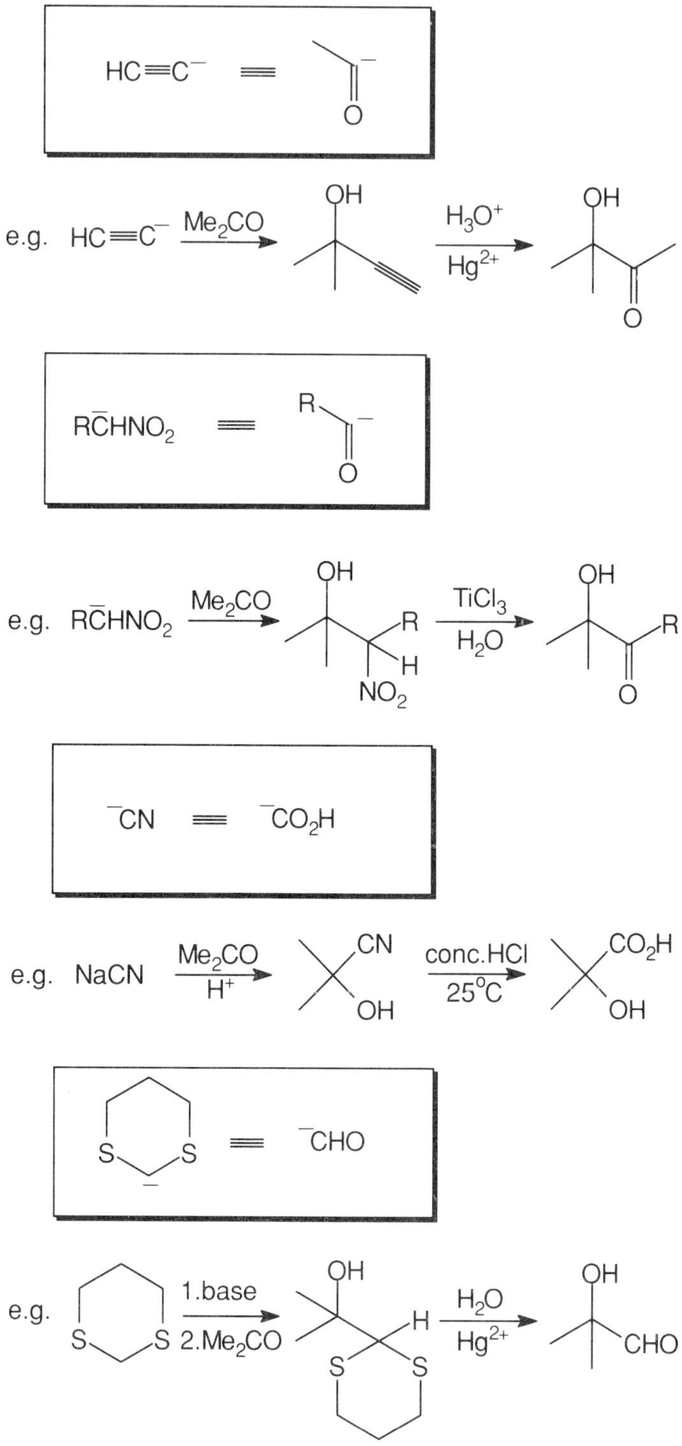

$$\text{(1,3-dithiane)} \quad\equiv\quad ^-CHO$$

e.g. (dithiane) $\xrightarrow[2.Me_2CO]{1.base}$ (dithiane adduct) $\xrightarrow[Hg^{2+}]{H_2O}$ (hydroxy aldehyde)

Scheme 2.4 Examples of the use of acyl anion equivalents

Benzoin condensation

The benzoin condensation is a dimerisation reaction of aldehydes. It involves treating the aldehyde (usually aromatic) with a catalytic amount of cyanide (eqn 2.9).

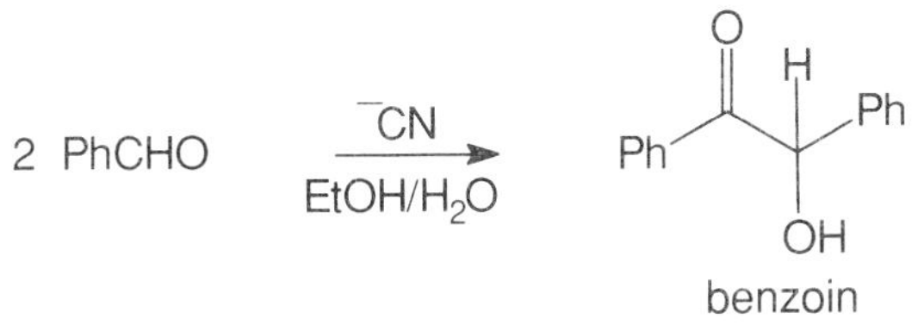

$$2 \ PhCHO \xrightarrow[\text{EtOH/H}_2\text{O}]{^-\text{CN}}$$

benzoin
(2.9)

The mechanism of the benzoin condensation is thought to involve the formation of a carbanion which then reacts with the second molecule of the aldehyde (scheme 2.5).

The benzoin condensation succeeds because the cyanide group can do three things :-
(1) ⁻CN is a good nucleophile
(2) CN group stabilises α-carbanions
(3) ⁻CN is a good leaving group.

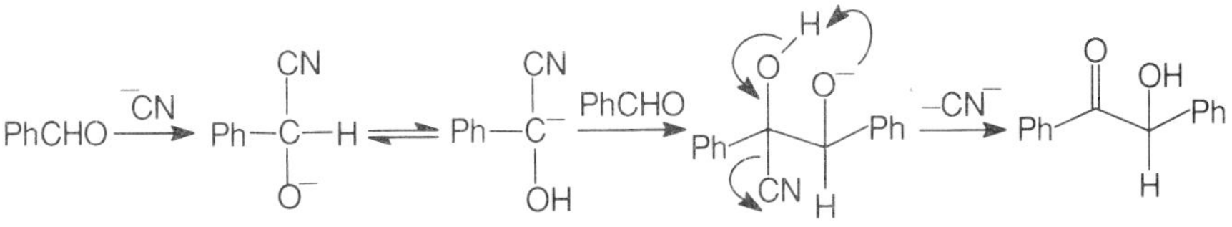

Scheme 2.5

The carbanion which is generated in the reaction is in fact acting as an acyl anion equivalent since this unit subsequently eliminates HCN to generate the carbonyl group in benzoin (eqn 2.10).

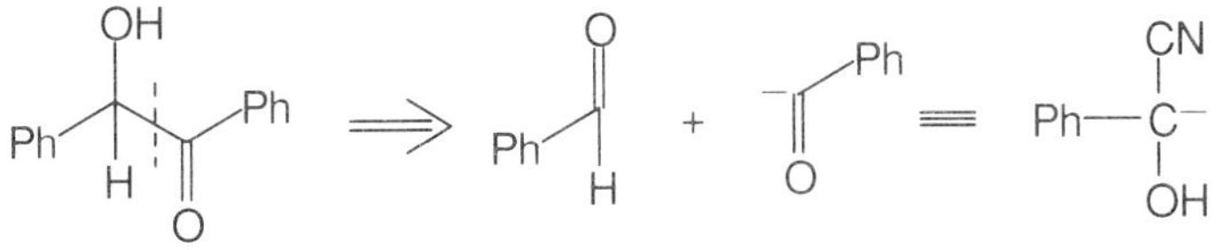

(2.10)

Acyloin condensation

The acyloin condensation involves a reductive dimerisation of an ester (eqn 2.11).

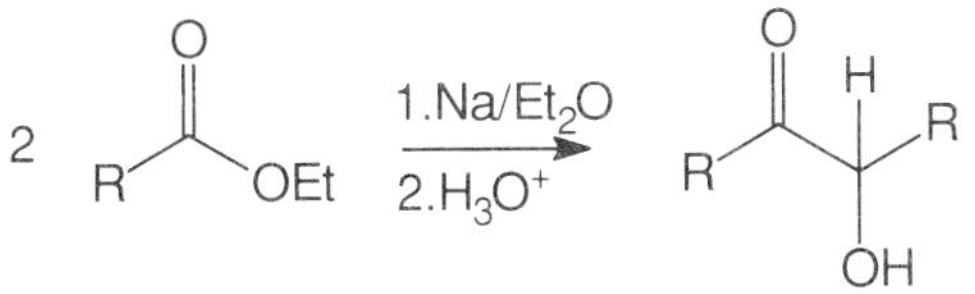

$$2 \quad \underset{R}{\overset{O}{\parallel}}\text{C} \underset{OEt}{} \xrightarrow[\text{2.H}_3\text{O}^+]{1.\text{Na/Et}_2\text{O}} $$

(2.11)

The mechanism of this reaction is similar to that of pinacol formation in that it involves electron transfer from the metal to form a radical anion which dimerises to give a diketone. Not surprisingly, this is also reduced under the reaction conditions to give, after quenching with water, an enediol which tautomerises to give the α-hydroxyketone (scheme 2.6).

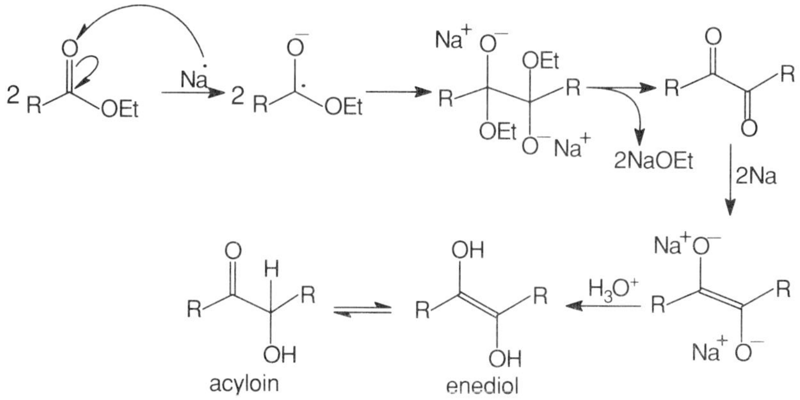

Scheme 2.6

The acyloin condensation is particularly useful for making cyclic α-hydroxyketones from diesters (eqn 2.12). The two ends of the diester are held together, and react together, on the metal surface.

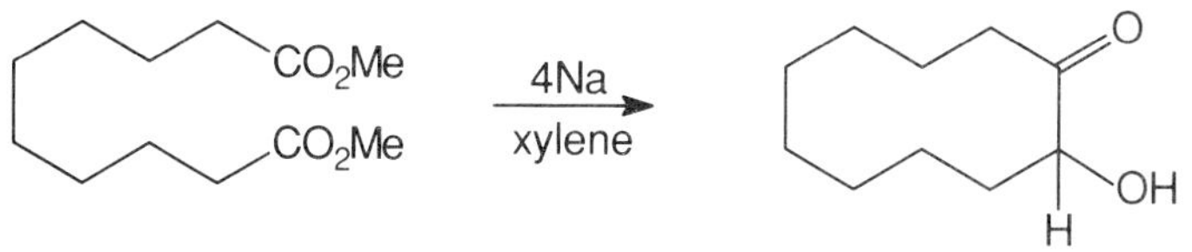

$$(2.12)$$

However a number of refinements have now been introduced which also make it possible to obtain good yields of acyclic and small ring cyclic α-hydroxyketones using the acyloin condensation. Thus, one of the by-products of the reaction is sodium ethoxide which is capable of bringing about several base-catalysed side-reactions including the Dieckmann condensation (see section 2.3). By adding four equivalents of trimethylsilyl chloride to the reaction, the ethoxide is trapped as Me_3SiOEt and the product dianion is converted into its bis-trimethylsilyl ether. Base-catalysed side-reactions are thereby precluded and the product can be readily regenerated by acid hydrolysis (e.g. eqns 2.13 and 2.14).

$TMSCl = Me_3SiCl$ $ArCH_2CO_2Et$

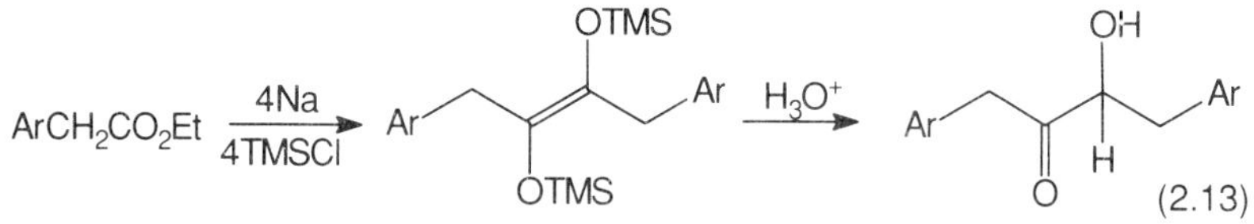

$$(2.13)$$

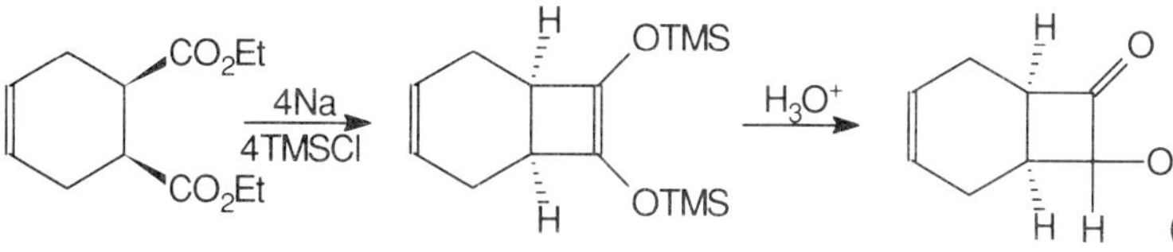

$$(2.14)$$

From α-halocarbonyl compounds

α-Halocarbonyl compounds afford direct access to 1,2-bifunctional compounds. For example, α-amino acids can be prepared from the corresponding α-halo acid which in turn can be prepared by halogenation of the corresponding carboxylic acid (scheme 2.7). Although ammonia can be used for the conversion of simple α-halo acids to α-amino acids, potassium phthalimide (followed by treatment with hydrazine) or azide ion (followed by reduction) are superior reagents (see J. Jones, *Amino acid and peptide synthesis*; in this series).

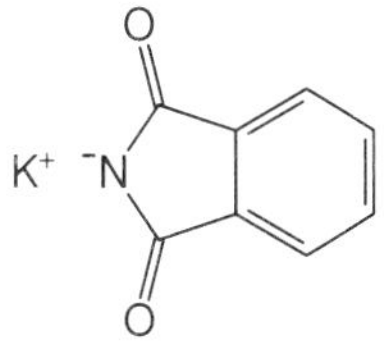

potassium phthalimide

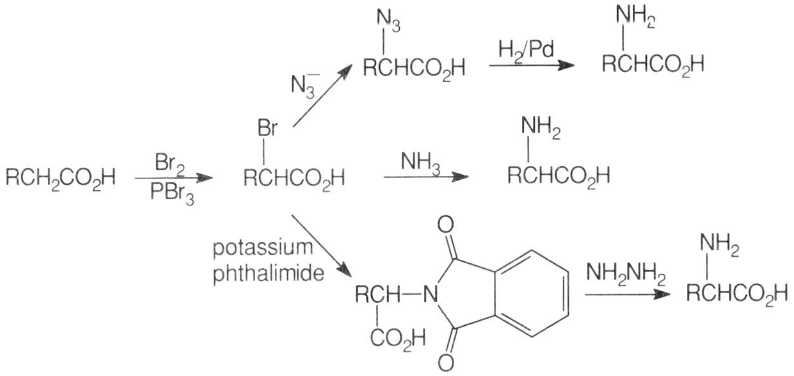

Scheme 2.7

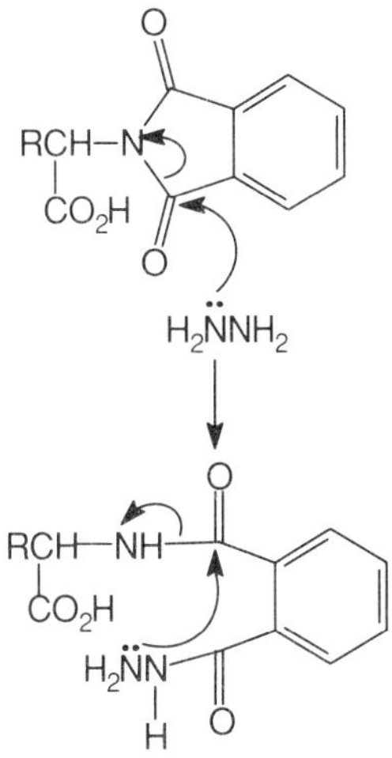

More generally, aldehydes and ketones can be halogenated at the α-position by reaction with a halogen in acidic solution (eqn 2.15). This reaction involves acid-catalysed enolisation of the aldehyde or ketone followed by reaction of the enol with the halogen. α-Halocarbonyl compounds are extremely versatile precursors for the synthesis of bifunctional molecules since the halogen substituent readily undergoes nucleophilic displacement by an S_N2 mechanism. α-Halocarbonyl compounds can also be used to prepare αβ-unsaturated carbonyl compounds by elimination, and they are also used in the Reformatsky and Wittig reactions (see section 2.3).

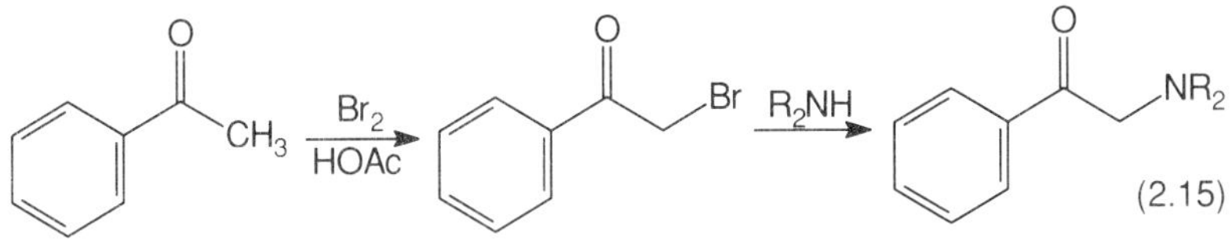

Mechanism of removal of phthalimido protecting group.

$$(2.15)$$

Oxidation

1,2-Dicarbonyl compounds are readily prepared by mild oxidation of α-hydroxycarbonyl compounds (eqns 2.16 and 2.17) and indeed, in some cases, by oxidation of unsubstituted carbonyl compounds (eqn 2.18).

$$(2.16)$$

$$(2.17)$$

$$(2.18)$$

The mechanism of selenium dioxide oxidation involves the formation of an enolate derivative which activates the α-carbon atom (eqn 2.19). Related methods involve α-bromination or nitrosation of carbonyl compounds (eqns 2.20 and 2.21).

$$(2.19)$$

$$(2.20)$$

$$(2.21)$$

2.3   Preparation of 1,3-bifunctional compounds

A retro-synthetic analysis of 1,3-bifunctional compounds is shown in scheme 2.8. Thus, β-hydroxycarbonyl compounds may be prepared using an aldol reaction, or by using the Reformatsky reaction, while β-dicarbonyl compounds are usually prepared by means of the Claisen condensation. αβ-Unsaturated carbonyl compounds can be prepared using the aldol condensation, or by using one of a number of similar reactions (e.g. Stobbe, Knoevenagel, Wittig reactions) which involve addition of a carbanion to an aldehyde or ketone. They can also be

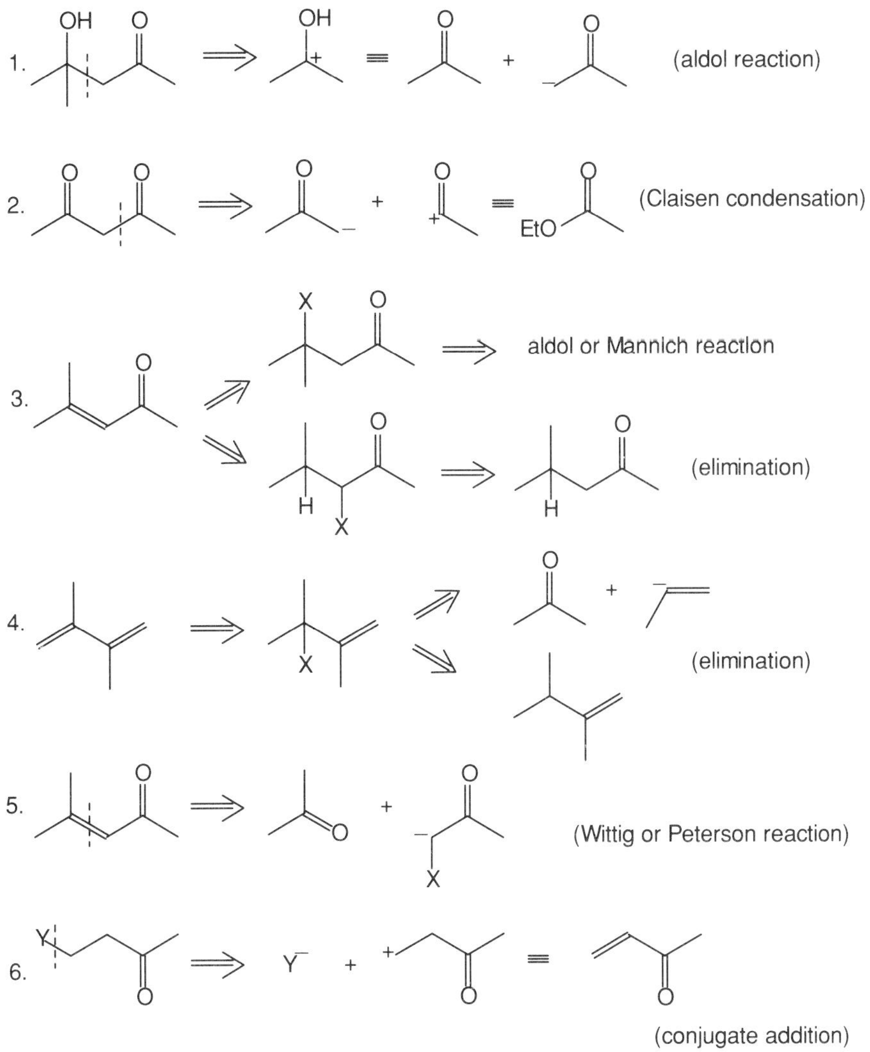

Scheme 2.8 Retro-synthetic analysis of 1,3-bifunctional compounds

prepared by an extension of the Mannich reaction, by elimination from an α-substituted carbonyl compound, or by the oxidation of an allylic alcohol by MnO_2 (see section 9.1). Many β-substituted carbonyl compounds can be obtained by conjugate addition to αβ-unsaturated carbonyl compounds (see section 7.2).

Aldol reaction

The aldol reaction in its simplest form involves the base-catalysed (or occasionally acid-catalysed) dimerisation of aldehydes or ketones (eqn 2.22).

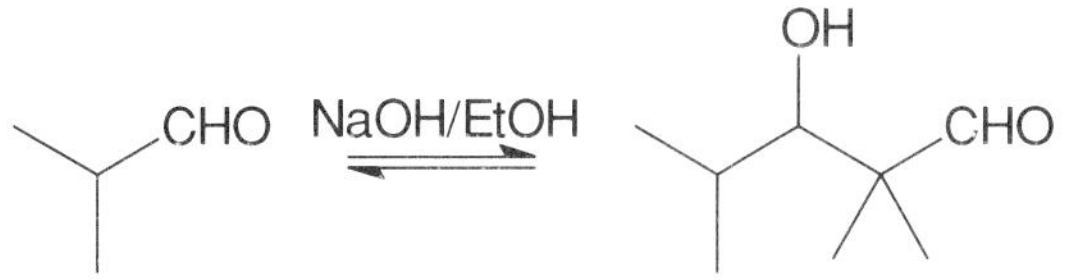

$$\text{(2.22)}$$

The reaction depends upon the acidity of the hydrogens α to the carbonyl group and the mechanism involves nucleophilic addition by the enolate anion derived from one molecule of the aldehyde (or ketone) to the carbonyl group of a second molecule (scheme 2.9).

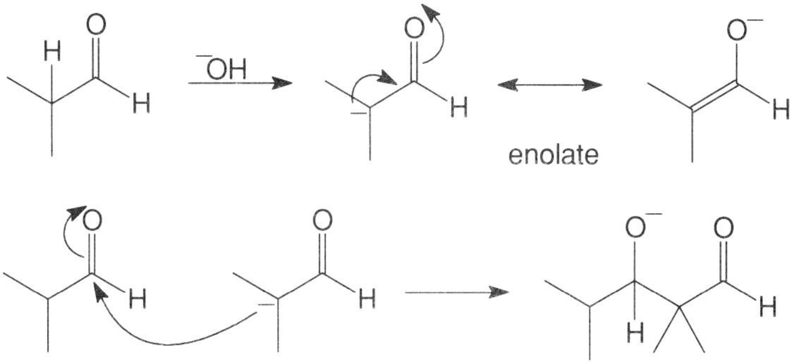

enolate

Scheme 2.9

Ketones undergo aldol reactions more reluctantly than aldehydes because the equilibrium in this case tends to favour the starting materials (eqn 2.23). Furthermore, since the aldol reaction is reversible β-hydroxycarbonyl compounds can in some cases undergo retro-aldol reactions. Note also that aldol products readily undergo dehydration to $\alpha\beta$-unsaturated carbonyl compounds, sometimes even under the conditions of the aldol reaction (see later).

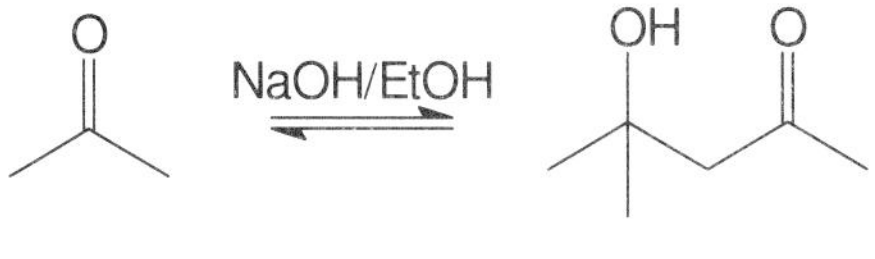

$$\text{(2.23)}$$

The aldol reaction is nevertheless an attractive method for making β-hydroxycarbonyl compounds, although until recently it had found only limited use in organic synthesis. One reason for this is that a mixed aldol reaction between two different aldehydes, or an aldehyde and a ketone, will give a mixture of products, unless one of the components has no α-hydrogens (e.g. eqn 2.24). A second reason is that a mixture of diastereoisomers is often produced (see below).

$$(2.24)$$

However these difficulties have now been largely overcome. Thus, a number of methods have been developed for bringing about aldol reactions between two different carbonyl compounds. Most of these involve making a preformed enolate or enol ether of one of the components (see section 2.5). The best results have been obtained by using lithium and boron enolates. For example, the use of a strong non-nucleophilic base such as lithium diisopropylamide (LDA) to generate the lithium enolate provides an effective method for bringing about mixed aldol reactions involving ketone enolates (eqn 2.25).

lithium diisopropyl-
amide (LDA)

$$(2.25)$$

An additional complication which is encountered for unsymmetrical ketones is the possibility of generating two different enolate derivatives, and hence obtaining isomeric products. Fortunately, the regioselectivity of the aldol reaction can usually be controlled by an appropriate choice of reaction conditions (scheme 2.10).

Scheme 2.10

The former procedure (low temperature, sterically hindered base) leads to the formation of the kinetically preferred enolate since the less hindered α-proton is abstracted more rapidly, while the latter procedure (higher temperature, equilibrating conditions) gives mainly the thermodynamically preferred (more stable) enolate.

The stereochemical course of the aldol reaction also depends upon whether it is carried out under kinetic or thermodynamic conditions. In the 'kinetic' reaction it is found that Z-enolates give mainly the *syn* aldol while E-enolates give mainly the *anti* aldol. The reactions are believed to proceed by way of a chair-like six-membered transition state in which the metal atom is bonded to the oxygen atoms of both the aldehyde and the enolate (scheme 2.11). By controlling the geometry of the enolate it is therefore possible to control the stereochemistry of the product. The superiority of

More substituted alkenes have a lower ΔH_f than less substituted ones.

The terms *syn* and *anti* are used to designate the two diastereoisomers which have two groups on the same or opposite sides of the molecule when it is represented as a zigzag conformation.

boron enolates is attributed to the shortness of the B–C and B–O bonds, which gives rise to a tighter transition state in which steric effects are enhanced (see section 2.5).

In contrast to reactions effected under kinetic conditions, those brought about under equilibrating conditions give mainly the *anti* aldol. Thus equilibration can sometimes be used to achieve *anti* stereoselection.

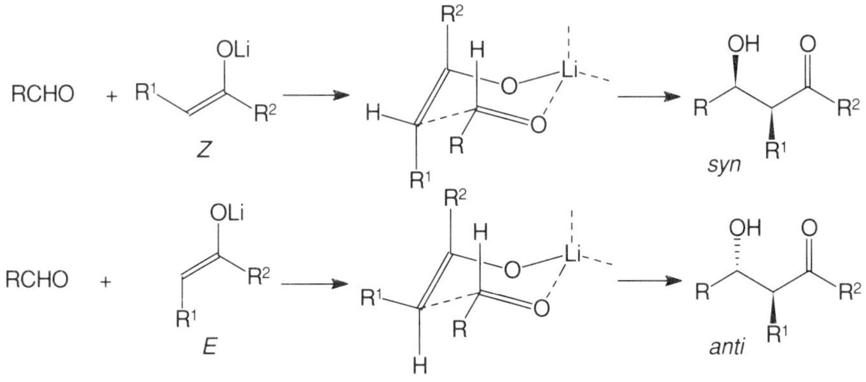

Scheme 2.11

Enolates derived from esters and amides can also be used in aldol-like reactions (e.g. eqn 2.26).

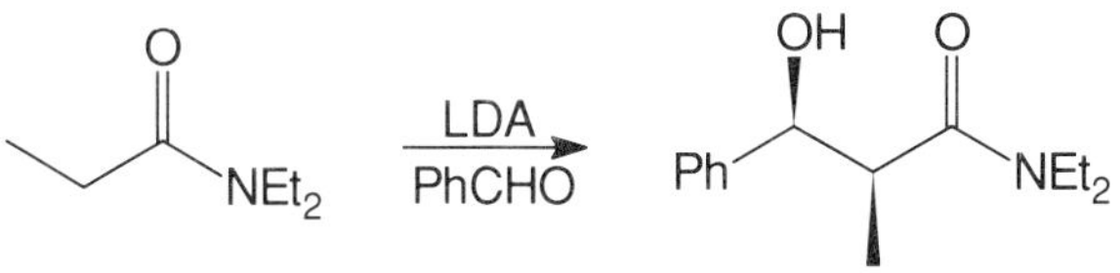

$$(2.26)$$

Reformatsky reaction

The Reformatsky reaction involves the formation of an organozinc reagent from an α-bromoester, and its addition to an aldehyde or ketone. Organozinc reagents are less reactive than lithium or magnesium reagents with the result that, although they react with aldehydes and ketones, they do not react with esters and therefore do not react with themselves. It is unclear whether the actual reagent is an alkyl zinc derivative or a zinc enolate (see eqn 2.27). In any event the product of the reaction is a β-hydroxyester, although this can easily be dehydrated to give the corresponding $\alpha\beta$-unsaturated compound (see later).

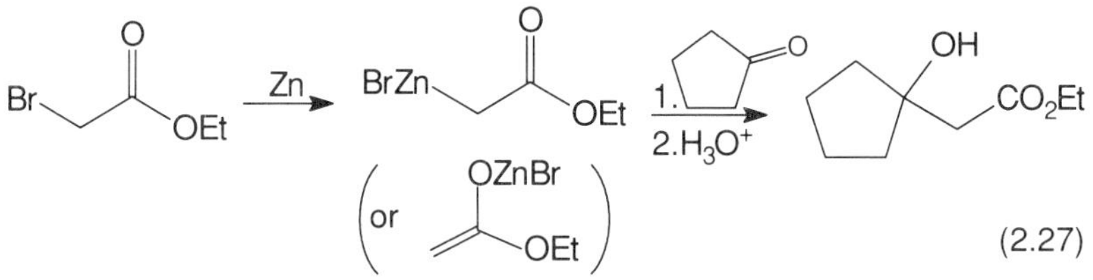

$$(2.27)$$

Claisen condensation

The Claisen condensation involves the combination of two molecules of an ester with the elimination of an alcohol (eqn 2.28).

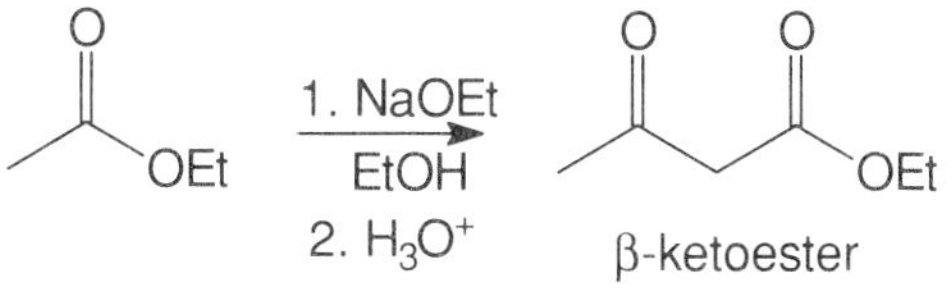

$$\text{(2.28)}$$

β-ketoester

Like the aldol reaction, the mechanism depends upon the acidity of the hydrogens α to a carbonyl group. It involves nucleophilic attack by an enolate anion derived from one molecule of the ester on a second ester molecule (scheme 2.12).

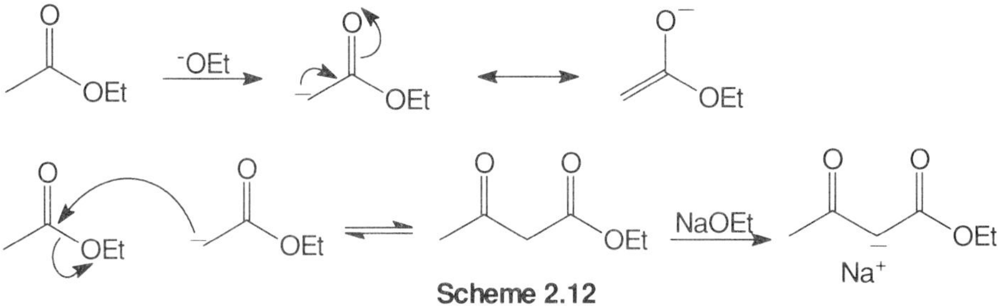

Scheme 2.12

The reaction is reversible and the equilibrium is shifted to the right by deprotonation of the product. Indeed the reaction fails if deprotonation of the product is not possible, unless a much stronger base is used (eqn 2.29). Furthermore, if 2,2-disubstituted β-ketoesters are treated with base a *retro* Claisen reaction occurs (see section 6.2).

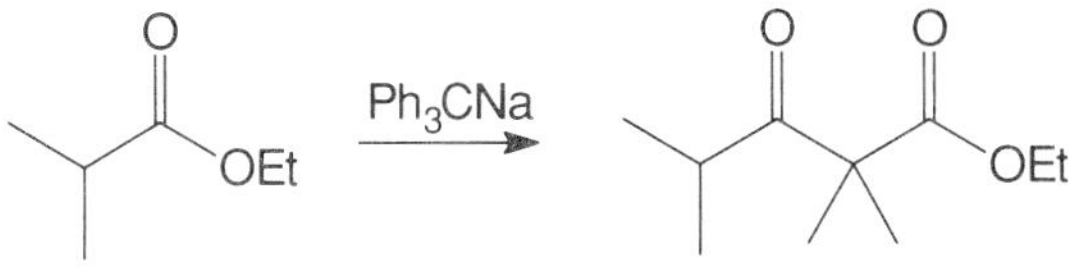

$$\text{(2.29)}$$

Mixed Claisen condensations usually yield a mixture of products unless one of the components has no α hydrogens (e.g. eqns 2.30 and 2.31).

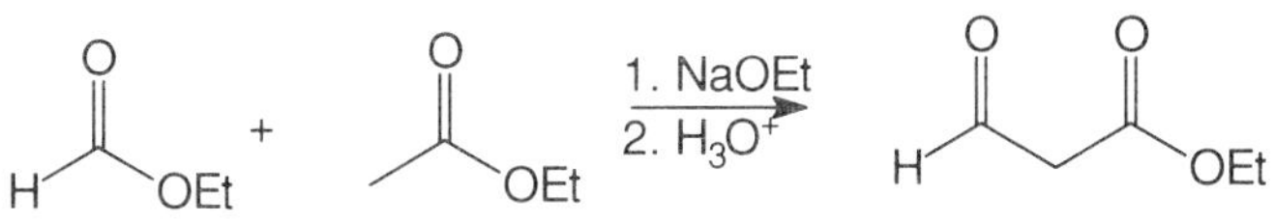

$$\text{(2.30)}$$

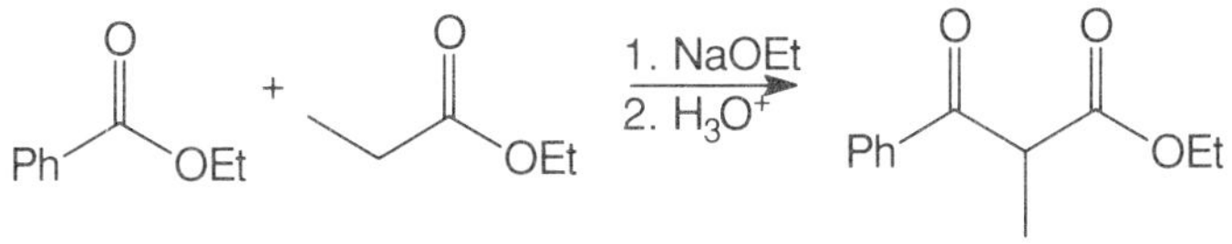

$$\text{(2.31)}$$

The reaction of a ketone enolate with an ester (preferably lacking α hydrogens) can be used to prepare β-diketones (eqn 2.32).

$$\text{(2.32)}$$

An intramolecular version of the Claisen condensation (known as the Dieckmann condensation) can be used to prepare cyclic β-ketoesters (eqn 2.33).

$$\text{(2.33)}$$

Elimination

$\alpha\beta$-Unsaturated carbonyl compounds can be prepared by the aldol condensation or by a number of similar reactions involving condensation of carbonyl compounds. Thus, unless the aldol reaction is carried out under carefully controlled conditions, or unless elimination is precluded, dehydration will readily occur (eqns 2.34 and 2.35).

Occurs by E1cB mechanism under basic conditions.

$$\text{(2.34)}$$

$$\text{(2.35)}$$

Certainly treatment of a β-hydroxycarbonyl compound with acid readily leads to elimination to give the $\alpha\beta$-unsaturated product (eqns 2.36 and 2.37).

Occurs by E1 or E2 mechanism under acidic conditions.

$$\text{(2.36)}$$

$$\text{(2.37)}$$

α-Bromoketones can also be dehydrobrominated by base to give $\alpha\beta$-unsaturated ketones (eqn 2.38).

The halogenation reaction involves the enol tautomer of the ketone. The more substituted (more stable) enol is obtained.

$$\text{(2.38)}$$

Alternatively, the introduction of a sulphoxide or selenoxide group provides an efficient method for preparing unsaturated carbonyl compounds by elimination. Sulphoxides and selenoxides with a β-hydrogen atom undergo a concerted *syn* elimination on heating to form alkenes (eqn 2.39). Since they are readily obtained by oxidation of sulphides and selenides respectively, this reaction provides a useful method for making αβ-unsaturated carbonyl compounds. Furthermore, the procedure has a number of advantages over the bromination-dehydrobromination sequence illustrated above, in that it takes place under comparatively mild conditions, and it can be carried out in the presence of other functional groups, such as carbon-carbon double bonds, which might be affected by bromination.

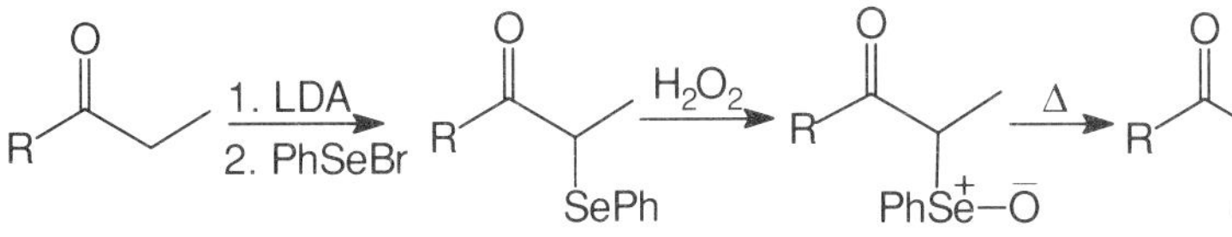

(2.39)

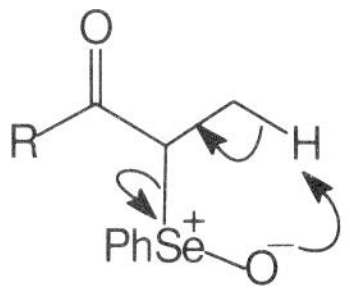

Another elimination procedure which can be used to prepare αβ-unsaturated carbonyl compounds makes use of the Mannich reaction (eqn 2.40). This involves the reaction of the enol of a carbonyl compound with an iminium ion generated from methanal and a dialkylamine. The β-dialkylaminocarbonyl compound is called a Mannich base and on heating it undergoes elimination to give the unsaturated carbonyl compound.

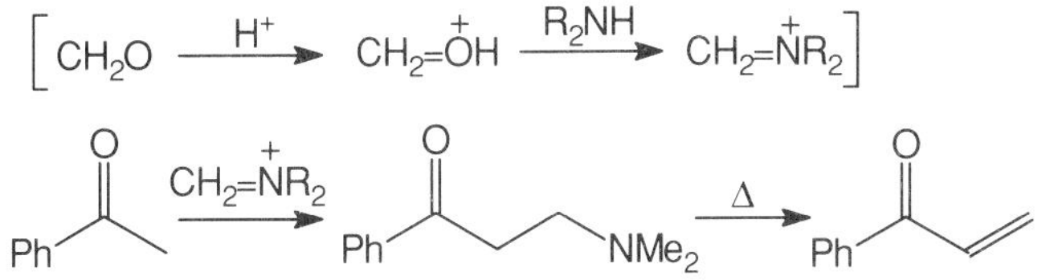

(2.40)

1,3-Dienes can also be prepared by elimination reactions starting from alkyl halides and alcohols. The reaction of an alkene with *N*-bromosuccin-imide (NBS) (eqn 2.41) involves abstraction of an allylic hydrogen atom, followed by reaction of the radical produced with a second molecule of NBS. This leads to a chain mechanism and results exclusively in substitution by bromine rather than addition to the carbon-carbon double bond. Notice that the acetylide anion can be used as a vinyl anion equivalent (eqn 2.42).

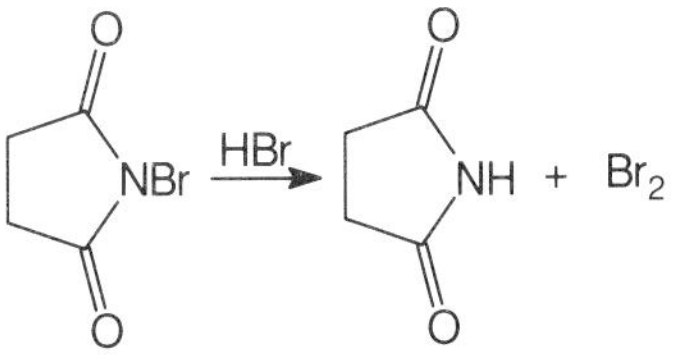

N-bromo-succinimide

NBS generates a low steady state concentration of bromine which is involved in the radical bromination reaction.
The allylic H is most readily abstracted since the allyl radical produced is stabilised by the π bond.

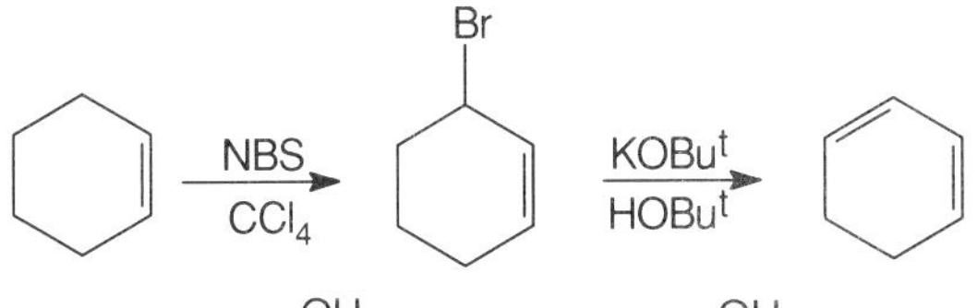

(2.41)

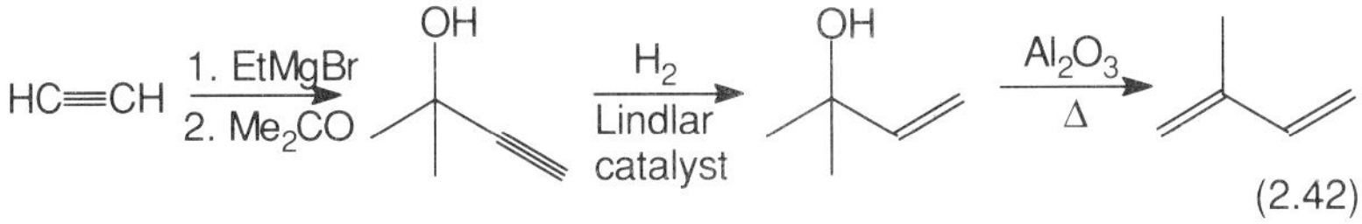

(2.42)

Wittig reaction

The Wittig reaction is a general method for the preparation of alkenes. It involves the formation of a phosphorus ylid or phosphorane by treating an alkyl halide with triphenylphosphine, followed by reaction with an aldehyde or ketone (eqn 2.43). The phosphorane is resonance-stabilised by overlap of the p orbital on carbon with one of the d orbitals on phosphorus.

A useful alternative to the conventional Wittig reaction involves the use of a phosphonate anion in place of the phosphorane (eqns 2.44). Phosphonate esters can be prepared from alkyl halides and triethylphosphite via the Arbuzov rearrangement. The carbanion obtained by treatment with base is more nucleophilic than the corresponding phosphorane and gives better yields than the conventional Wittig reaction. In addition, it has the practical advantage that the phosphate formed as a by-product is water soluble and can therefore be more easily separated from the product than triphenylphosphine oxide.

Mechanism of the Arbuzov reaction

$$(EtO)_2\overset{+}{P}\underset{O}{\overset{CH_2CO_2Et}{\diagdown}}Et \qquad X^-$$

$$(2.43)$$

$$(2.44)$$

In both cases the mechanism of the reaction involves the formation of a four membered intermediate followed by expulsion of triphenylphosphine oxide or phosphate (eqn 2.45). In the reactions of resonance-stabilised ylids and phosphonate anions the addition step is reversible with the result that the intermediate leading to the *E*-alkene predominates.

$$(2.45)$$

Peterson olefination

An alternative method of alkene preparation which has some similarities with the Wittig reaction involves the elimination of trimethylsilanol from a β-hydroxysilane. It has the practical advantage that the by-product of the reaction, hexamethyldisiloxane, is volatile and much easier to remove from the product mixture than triphenylphosphine oxide. Furthermore, the stereochemical course of the reaction can be more easily controlled since it follows two different mechanistic pathways depending upon whether the elimination is carried out under acidic or basic conditions.

Under acidic conditions *anti* elimination predominates, whilst under basic conditions *syn* elimination occurs, probably by way of a four-membered cyclic intermediate which breaks down in a manner similar to the final step of the Wittig reaction. The reaction can be used for the preparation of αβ-unsaturated esters (eqn 2.46), and is an attractive alternative to the phosphonate modification of the Wittig reaction since the silicon reagents are more reactive than phosphonate anions. Even easily enolisable ketones which often give poor yields in the Wittig reaction react readily with the silicon reagents.

syn elimination

anti elimination

(2.46)

The β-hydroxysilanes required for the Peterson reaction may be prepared by reacting α-metallated silanes with aldehydes and ketones, by reduction or nucleophilic addition to β-ketosilanes, or by reaction of αβ-epoxysilanes with organocuprate reagents (see S. E. Thomas, *Organic synthesis: the roles of boron and silicon;* in this series).

2.4 Preparation of 1,4- and 1,5-bifunctional compounds

Scheme 2.13 shows a retro-synthetic analysis of 1,4- and 1,5-bifunctional compounds. Most of the reactions involve the use of bifunctional compounds as starting materials and they are therefore dealt with in more detail in later chapters. For example, 1,4-dicarbonyl compounds can be made by alkylation of enamines or β-ketoesters (see sections 8.1 and 6.2). Note particularly, however, the possibility of preparing bifunctional compounds by oxidative cleavage of alicyclic compounds and the preparation of cyclohexadienes and even enol ethers (section 2.5) by Birch reduction.

Oxidative cleavage

A useful source of 1,5- (and 1,6-) bifunctional compounds involves the oxidative cleavage of five- and six-membered alicyclic compounds. For example, oxidation of a cyclopentene derivative by ozonolysis (or by conversion to the diol followed by treatment with periodate—see section 4.4) gives a 1,5-dicarbonyl compound (eqn 2.47), while Baeyer-Villiger oxidation of a cyclopentanone gives a six-membered lactone which can be hydrolysed to the corresponding δ-hydroxyacid (eqn 2.48).

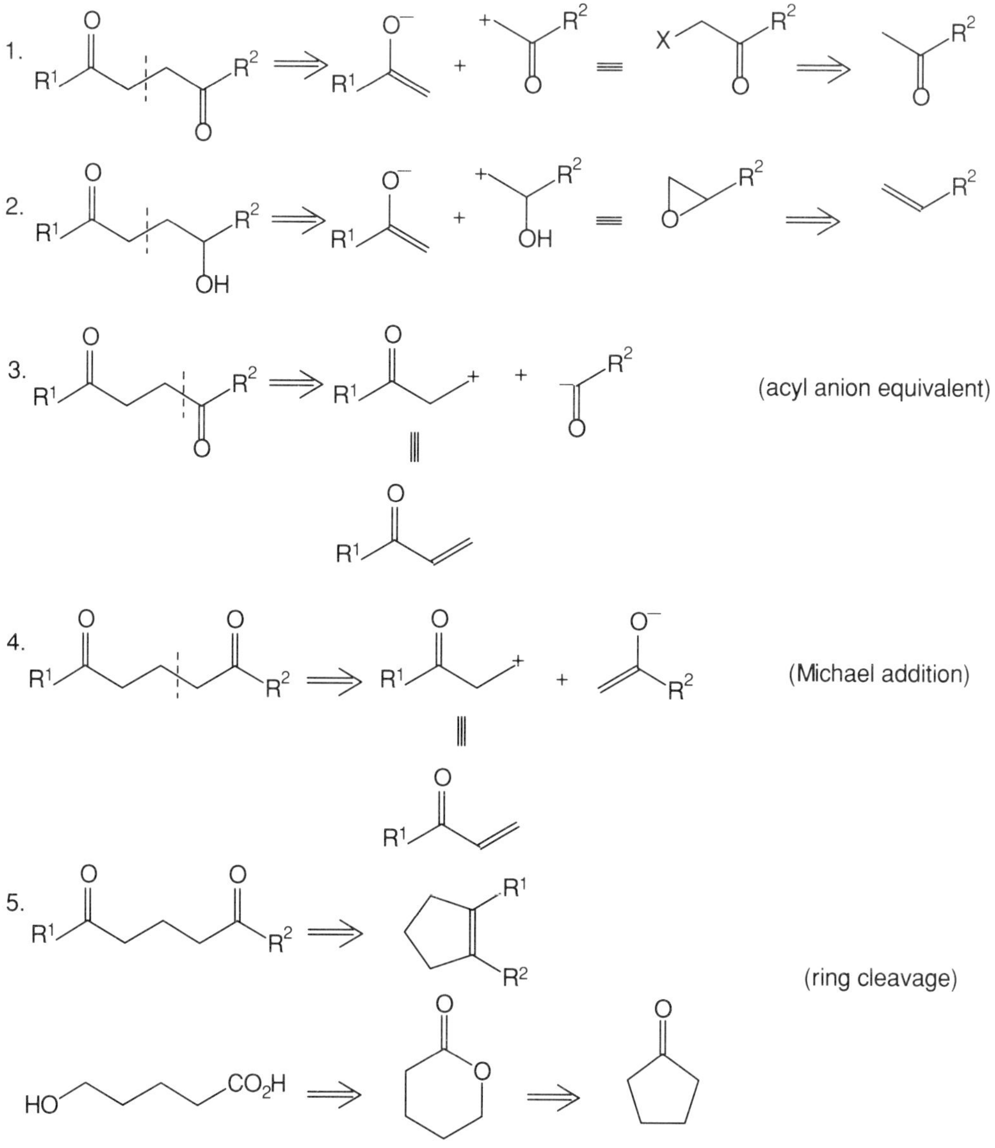

Scheme 2.13   Retro-synthetic analysis of 1,4- and 1,5-bifunctional compounds

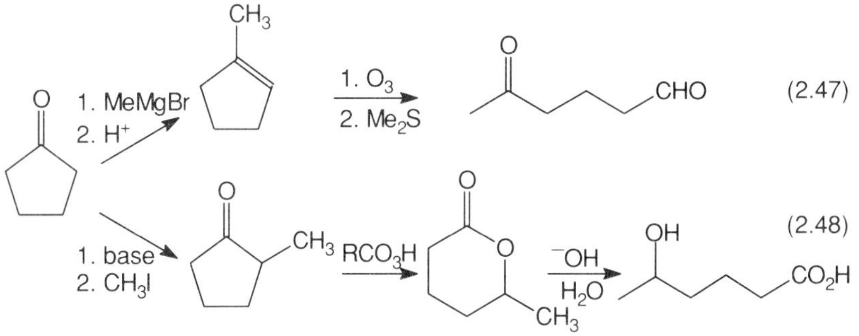

Bifunctional compounds   25

Mechanism of ozonolysis

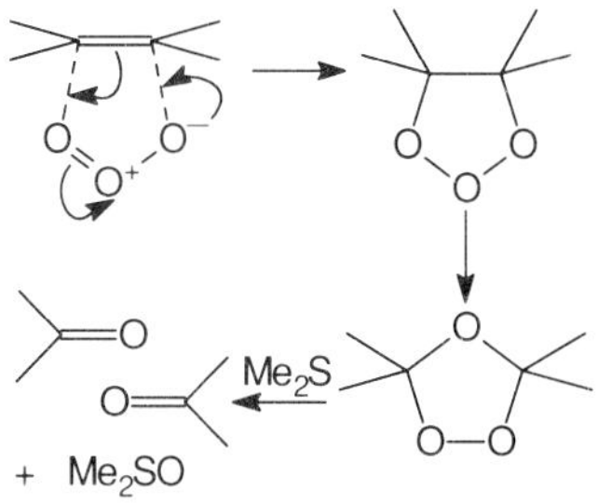

$$ (2.47) $$

$$ (2.48) $$

+ Me₂SO

Mechanism of the Baeyer–Villiger
reaction.

Birch reduction

1,4-Cyclohexadienes can be prepared by dissolving metal reduction of
aromatic compounds (eqn 2.49).

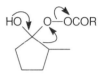

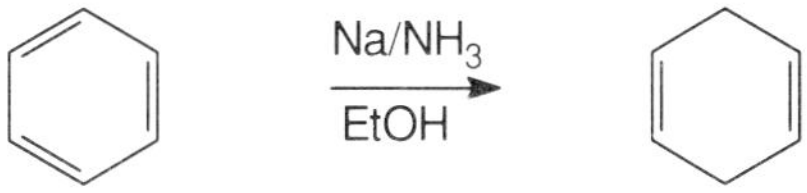

The more substituted carbon atom
migrates to the electron-deficient
centre.

$$ (2.49) $$

This reaction involves treating an arene (or other conjugated system–see
section 7.2) with an alkali metal (Na or Li) in liquid ammonia containing
an alcohol as a proton source (ammonia itself is not sufficiently acidic).
The mechanism of the reaction involves electron transfer from the metal
to give a radical anion which is then protonated by the alcohol. Repetition
of this process gives the 1,4-dihydrocompound (scheme 2.14) (see
M. Sainsbury, *Aromatic chemistry;* in this series).

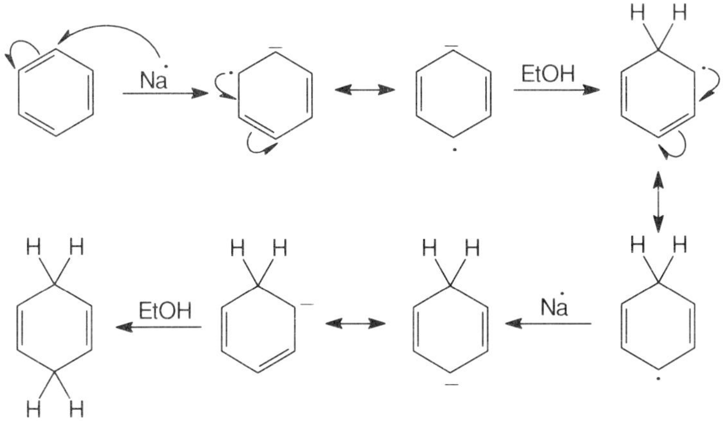

Scheme 2.14

2.5   Enamines, enol ethers, and enolates

Enamines are prepared by reacting an aldehyde or ketone with a secondary amine in the presence of an anhydrous acid catalyst. Typical secondary amines which are often used for this purpose are piperidine and morpholine (eqns 2.50 and 2.51). Some enol ethers can be prepared in a similar manner (eqn 2.52). Other enol ethers may be prepared from alkoxybenzenes by Birch reduction (eqn 2.53).

piperidine   morpholine

$$\text{(2.50)}$$

$$\text{(2.51)}$$

$$\xrightarrow[\text{H}^+]{\text{EtOH}} \qquad \text{(2.52)}$$

$$\xrightarrow[\text{Bu}^t\text{OH}]{\text{Na/NH}_3} \qquad \text{(2.53)}$$

One of the most useful groups of enol ethers are those derived from silicon. Silyl enol ethers can be isolated, purified and characterised using standard preparative procedures. They are usually prepared from the corresponding enolate anions by reaction with trimethylsilyl chloride (eqns 2.54 and 2.55). In the case of unsymmetrically substituted ketones two different silyl enol ethers can be prepared depending upon whether the enolate is generated under 'kinetic' or 'thermodynamic' conditions.

The dividing line between enol ethers and enolates is a narrow one, depending upon the position of the element attached to oxygen in the periodic table. Boron enolates can be prepared from ketones by reaction with a dialkylboryl triflate (trifluoromethanesulphonate) in the presence of a tertiary base. Once again by careful choice of reaction conditions it is possible to prepare two isomeric boron enolates from unsymmetrical ketones. Dibutylboron triflate gives the 'kinetic' (less substituted) derivative, while 9-borabicyclo[3.3.1]nonanyl triflate gives the 'thermodynamic' (more substituted) one (eqns 2.56 and 2.57) (see chapter 8 for further discussion).

triflic acid

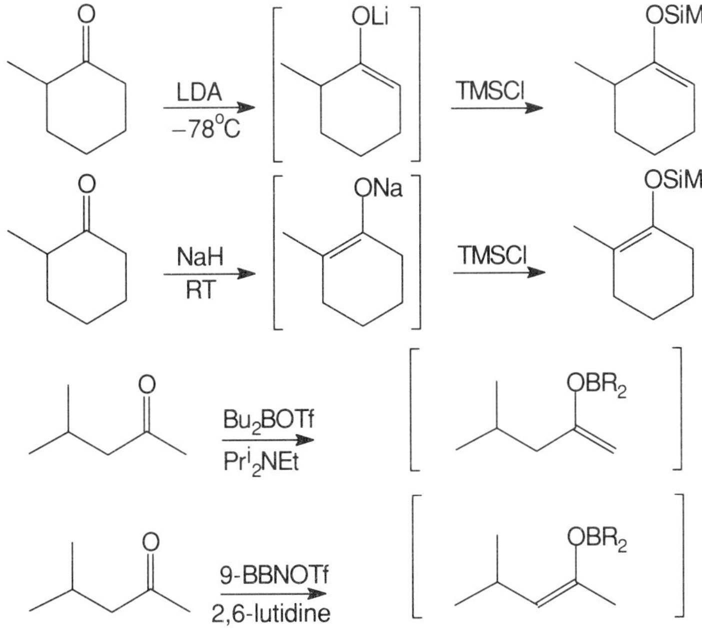

$$\text{(2.54)}$$

$$\text{(2.55)}$$

$$\text{(2.56)}$$

$$\text{(2.57)}$$

BBNOTf =

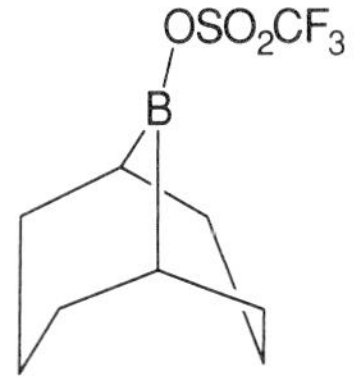

9-borabicyclo[3.3.1]nonanyl
trifluoromethanesulphonate

2,6-lutidine =

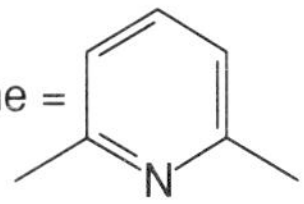

2,6-dimethylpyridine

2.6 Allenes, ketenes, and heterocumulenes

Allenes are often prepared by isomerisation of alkynes (eqn 2.58).

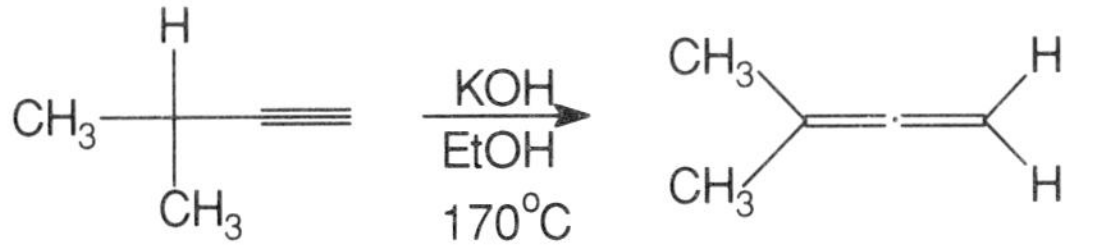

$$\text{(2.58)}$$

Ketenes are usually prepared by elimination of HX or X_2 from an acyl halide or α-haloacyl halide (eqns 2.59 and 2.60), or from a diazoketone by the Wolff rearrangement (eqn 2.61).

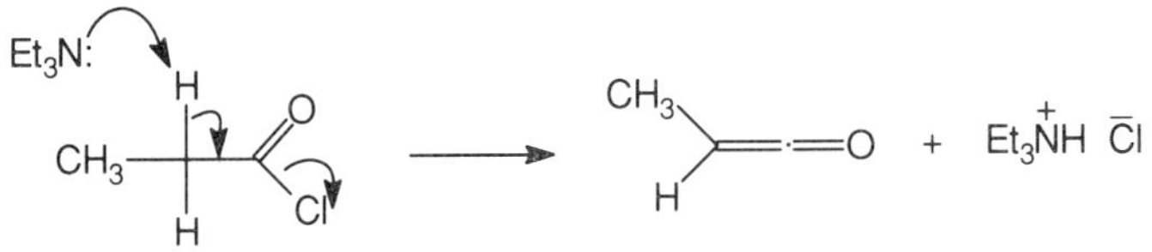

$$\text{(2.59)}$$

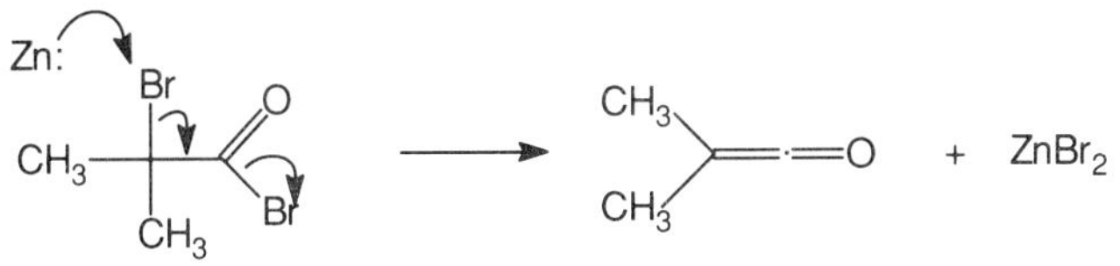

$$\text{(2.60)}$$

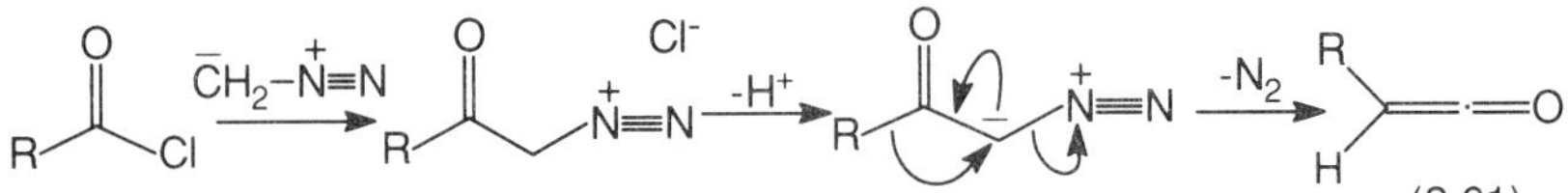

$$\text{(2.61)}$$

Ketene itself is prepared by pyrolysis of acetone (eqn 2.62).

$$\text{(2.62)}$$

Isocyanates are produced as intermediates in the Hoffmann rearrangement of amides (eqn 2.63) and in the Curtius rearrangement of acyl azides (eqn 2.64). They can also be prepared by reacting alkyl halides with sodium cyanate (eqn 2.65) or by treating an amine with phosgene (eqn 2.66).

$$\text{(2.63)}$$

$$\text{(2.64)}$$

$$RBr \ + \ Na^+ \ ^-NCO \longrightarrow R-N=C=O \qquad \text{(2.65)}$$

$$PhNH_2 \xrightarrow{COCl_2} PhNHCOCl \xrightarrow{\Delta} Ph-N=C=O \qquad \text{(2.66)}$$

Problems

Suggest methods for carrying out each of the following transformations. More than one step may be required in each case.

1.

2.

3.

4.

3 Reactions of dienes

3.1 Structure and properties of dienes

In *allenes* (1,2-dienes) two carbon–carbon double bonds share a common carbon atom. This carbon atom is sp^1 hybridised and the molecular orbitals of the two double bonds are therefore at right angles to one another. As a result the two ends of the molecule are held in perpendicular planes and some allenes are indeed chiral (fig. 3.1).

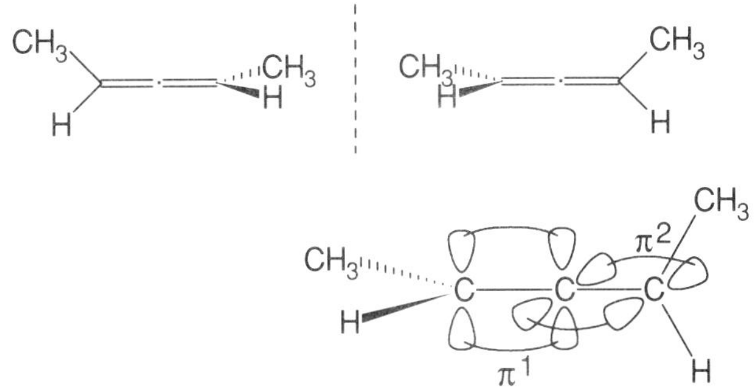

Figure 3.1

Allenes are isomeric with alkynes and the two classes of compounds can often be interconverted (see section 2.6). The reactions of allenes are similar to those of alkenes and alkynes.

A second very important class of dienes is that in which two doubly bonded carbon atoms are separated by a single bond. These compounds are called *conjugated dienes* because the unhybridised 2p atomic orbitals on all four carbon atoms interact to give π orbitals which embrace the entire four-carbon system (fig. 3.2).

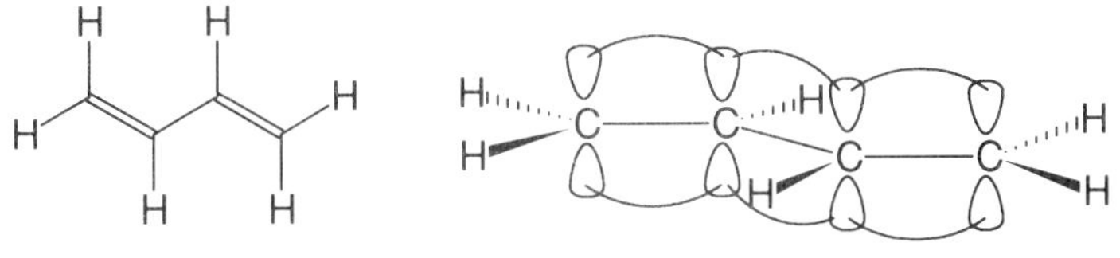

Figure 3.2

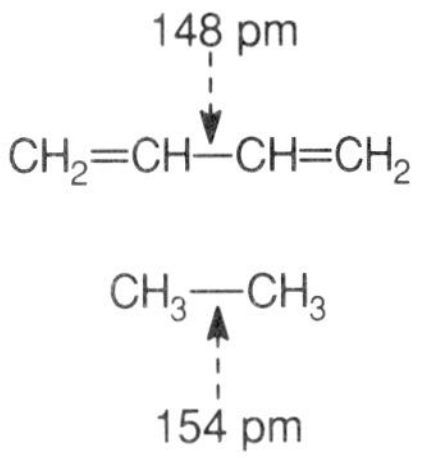

The C2–C3 bond in 1,3-butadiene is in fact slightly shorter than the carbon–carbon single bond in, for example, ethane. This difference can be explained in terms of the different hybridisation of the orbitals involved. Thus, since a C–H bond involving an sp^2 hybrid orbital is shorter than one involving an sp^3 hybrid orbital, then a C–C single bond involving sp^2

$$CH_2{=}CH{-}CH{=}CH_2 \quad 148 \text{ pm}$$

$$CH_3{-}CH_3 \quad 154 \text{ pm}$$

hybrid orbitals (as in butadiene) would be expected to be shorter than one involving sp^3 hybrid orbitals (as in ethane).

Certainly any overlap of the unhybridised 2p atomic orbitals on carbon atoms 2 and 3 of butadiene can only occur when the molecule is planar. This imposes a small barrier to rotation about the central C–C bond, so that the preferred conformations of 1,3-butadiene are those in which all four carbon atoms are coplanar (eqn 3.1). Nevertheless, at room temperature the two conformers are easily interconverted.

$$\text{(3.1)}$$

The molecular orbital picture of conjugated dienes is supported by their characteristic UV spectra. Indeed compounds containing extended conjugation are highly coloured and many natural pigments are of this type (see section 3.4).

3.2   Addition reactions of conjugate dienes

Conjugated dienes react with the same kinds of reagent as alkenes. However in some cases unexpected products are obtained. For example, the reaction of one mole of 1,3-butadiene with one mole of hydrogen bromide gives two products, one corresponding to 1,2-addition, the other corresponding to 1,4-addition (eqn 3.2).

$$\text{(3.2)}$$

Furthermore, the proportions of the two products vary depending upon the reaction temperature (eqn 3.3). At low temperatures the kinetic product predominates, whilst at higher temperatures the thermodynamic product is favoured. This is because the 1,2-adduct is formed fastest (lowest energy barrier) and is therefore the major product formed under mild conditions. However, the 1,4-adduct is more stable and, since the steps leading to both products are reversible, this is the major product obtained when sufficient thermal energy is available to surmount the higher energy barrier leading to the 1,4-adduct.

$$\text{(3.3)}$$

A similar situation is observed in the reaction of 1,3-butadiene with bromine (eqn 3.4). Two products are obtained, one resulting from 1,2-addition and the other from 1,4-addition. The proportions of the two products obtained again depend upon the reaction conditions employed.

$$(3.4)$$

3.3 Cycloaddition reactions

Conjugated dienes undergo one reaction which simple alkenes do not undergo. This is the *Diels–Alder reaction* in which the diene reacts with a 'dienophile', which is usually an alkene or alkyne bearing one or more electron-withdrawing groups. This reaction is the prototype of a general class of concerted reactions, called *pericyclic reactions*, which involve redistribution of electrons around a cyclic transition state (e.g. eqn 3.5). There are three main types of pericyclic reactions. This particular type of reaction in which two σ bonds are formed between the ends of two π systems is called a *cycloaddition reaction*. The Diels–Alder reaction can be formally described as a [4 + 2] cycloaddition, indicating the number of participating electrons contributed by each component. It should be noted however that not all cycloaddition reactions are necessarily concerted, and in some cases they may involve diradical or dipolar intermediates.

Main types of pericyclic reaction:
1. cycloaddition reactions
2. electrocyclic reactions
3. sigmatropic shifts

$$(3.5)$$

The Diels–Alder reaction is an important method for the preparation of compounds containing six-membered rings. A number of features of the reaction are of special interest.

1. Concerted reactions such as the Diels–Alder reaction and other reactions in which ring closure is rapid relative to bond rotation, are invariably highly stereoselective. In contrast, reactions proceeding via a multi-step mechanism involving ring opened intermediates are usually less stereoselective. The Diels–Alder reaction is in fact stereospecific, since the

stereochemistry of the dienophile and the diene are retained in the product. Using the terminology used to describe such reactions, the Diels–Alder reaction is said to involve *suprafacial* (or *syn*) addition (as opposed to *antarafacial* or *anti* addition) to both of the reactants, i.e. the diene and the dienophile (fig. 3.3).

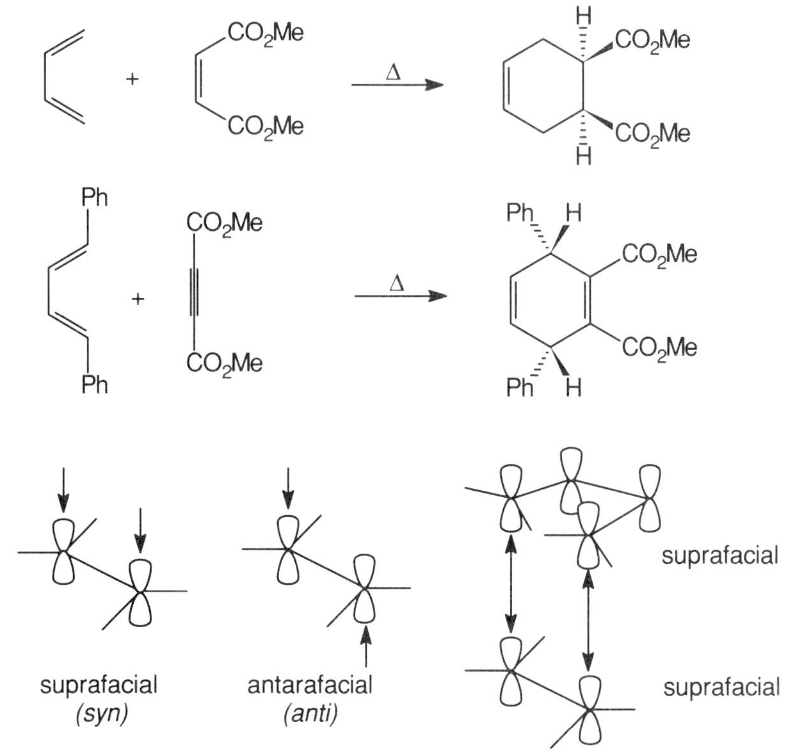

Figure 3.3

2. The reaction is also stereoselective in the sense that *exo* and *endo* products are obtained in unequal amounts. The *endo* (kinetic) product is usually the major product although the *exo* adduct is usually the thermodynamically more stable and may predominate if long reaction times and high temperatures are used. The transition state leading to the *endo* adduct is stabilised by through-space interactions known as secondary orbital interactions (fig. 3.4).

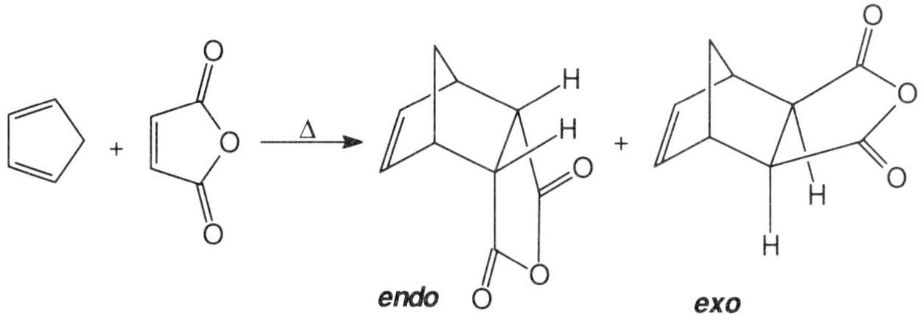

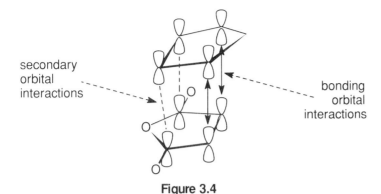

secondary orbital interactions

bonding orbital interactions

Figure 3.4

Figure 3.4 shows the orbital overlap in the transition state leading to the *endo* adduct.

3. The Diels–Alder reaction is also regioselective when unsymmetrical dienes and unsymmetrical dienophiles are involved (eqns 3.6 and 3.7). By using frontier molecular orbital theory it is in fact possible to rationalise the regiochemical preferences observed.

Frontier molecular orbital theory involves consideration of the interaction between the highest occupied m.o. (HOMO) of one reactant and the lowest unoccupied m.o. (LUMO) of the other.

major   minor   (3.6)

major   minor   (3.7)

4. Diels–Alder reactions can be catalysed by Lewis acids and by carrying out such reactions at low temperatures it is possible to greatly enhance both the regio- and stereoselectivity. Furthermore by employing a chiral catalyst it is now possible to bring about enantioselective Diels–Alder reactions, leading to non-racemic products.

3.4 Polyenes

Many natural products contain extended conjugated systems and some are highly coloured. Examples include the carotenoids lycopene, a red pigment in ripe tomatoes and watermelons, and β-carotene, an orange pigment present in many plants and first isolated from carrots.

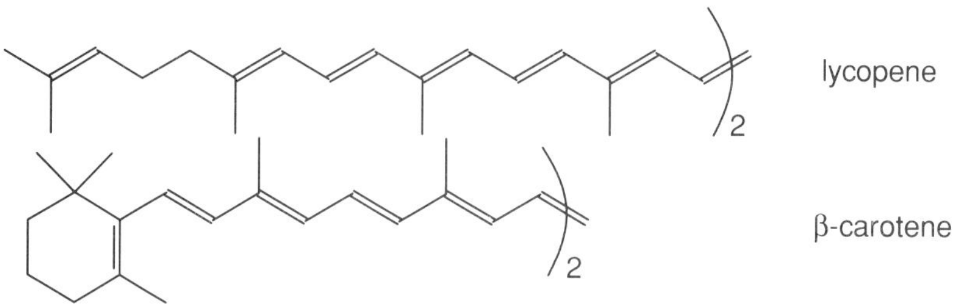

lycopene

β-carotene

β-Carotene and similar cyclic carotenoids can be cleaved in the liver to give vitamin A. Vitamin A is also closely related to retinal, a component of the photosensitive substance rhodopsin that is found in the retina of the eye. Rhodopsin is a red complex of 11-*cis*-retinal with the protein opsin which is converted into *trans*-retinal on absorption of light. Reduction of this aldehyde to *trans*-vitamin A followed by isomerisation in the dark gives 11-*cis*-vitamin A which is reoxidised to 11-*cis*-retinal (scheme 3.1).

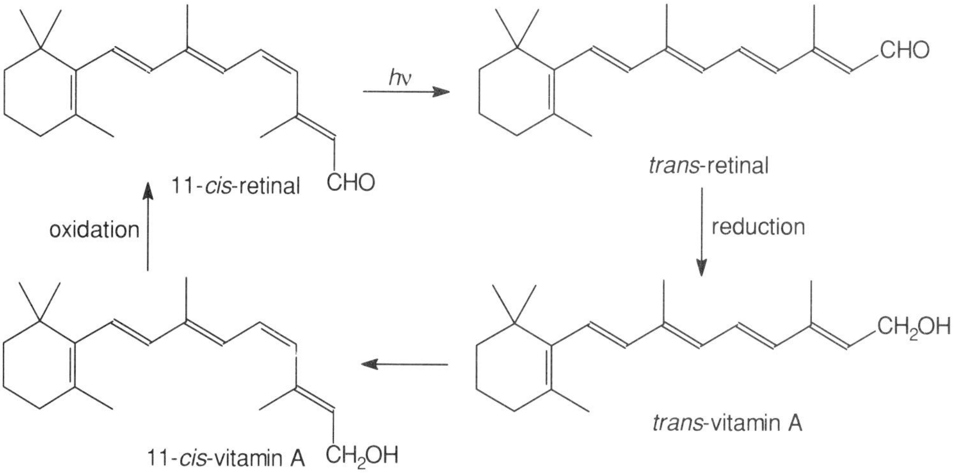

Scheme 3.1

3.5   Cope rearrangement of 1,5-dienes

1,5-Dienes undergo a concerted pericyclic rearrangement on heating. This particular reaction is known as the Cope rearrangement but it belongs to a general class of pericyclic reactions called *sigmatropic reactions* or sigmatropic shifts which involve the movement of a σ bond across one (or more) π systems. The Cope rearrangement is an example of a [3,3] sigmatropic reaction (eqn 3.8). It is believed to involve a chair-like six-membered transition state, and the stereochemical outcome can be rationalised using the general principles of conformational analysis, namely that conformations are preferred in which bulky groups adopt equatorial positions (scheme 3.2).

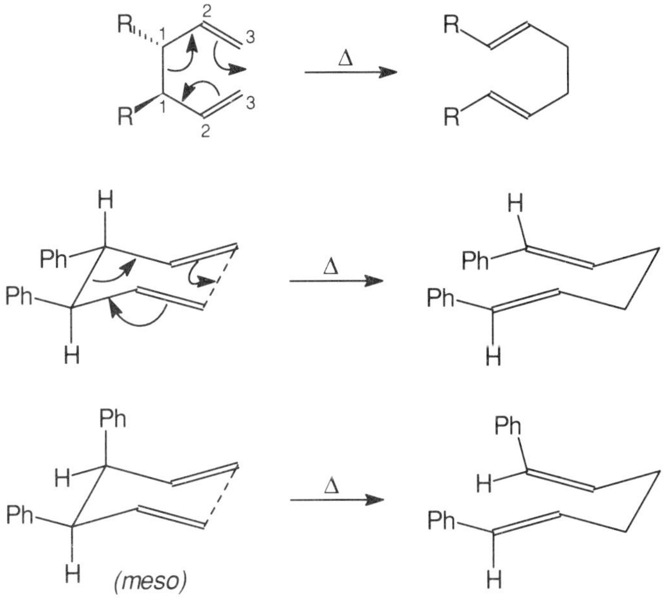

(3.8)

(E,E)

(E,Z)

Scheme 3.2

The so-called oxy-Cope rearrangement of 1,5-dienes having a hydroxyl substituent at C-3 provides a useful route to unsaturated aldehydes and ketones (eqn 3.9).

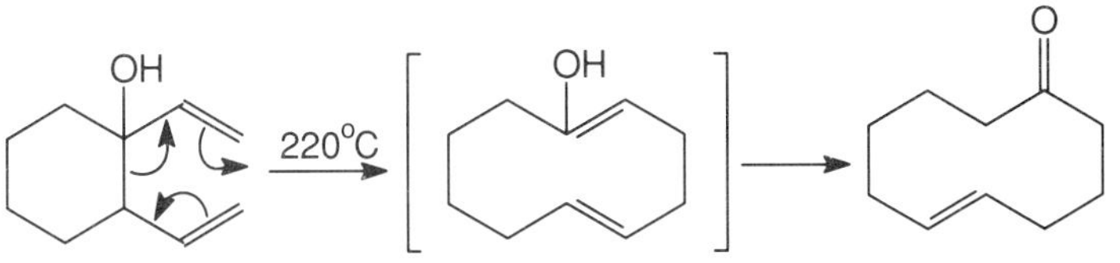

(3.9)

A disadvantage of the Cope rearrangement from a preparative stand-point is the high temperature required. However, very large increases in the rate are obtained by using the potassium alkoxides rather than the hydroxy compounds themselves (eqn 3.10).

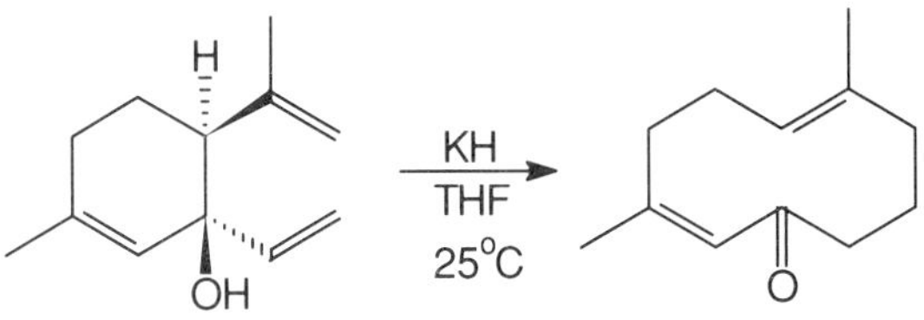

(3.10)

Problems

1. Suggest a structure for the intermediate **A** formed in the following sequence and suggest mechanisms for the reactions taking place.

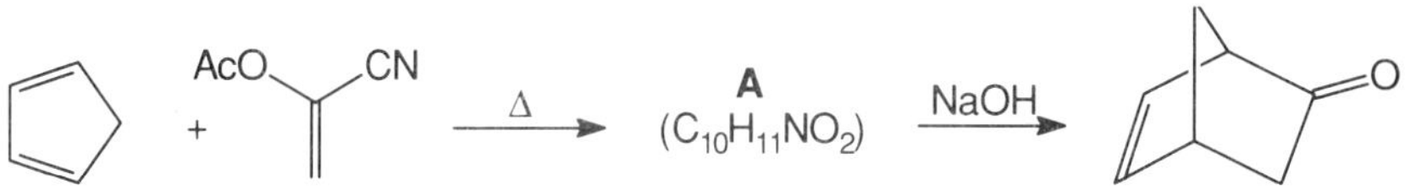

2. Suggest a mechanism for the following reaction.

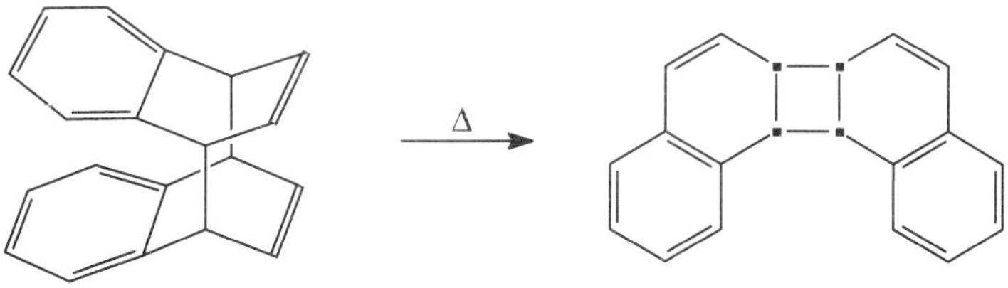

4 Reactions of diols

4.1 Pinacol–pinacolone rearrangement of 1,2-diols

1,2-Diols undergo acid-catalysed rearrangement to afford ketones (scheme 4.1). This is a general reaction of 1,2-diols and involves the generation of a carbocation intermediate followed by migration of an alkyl group to the electron-deficient centre. The driving force for the reaction is provided by the resonance stabilisation of the resulting carbocation (see L. M. Harwood, *Polar rearrangements;* in this series).

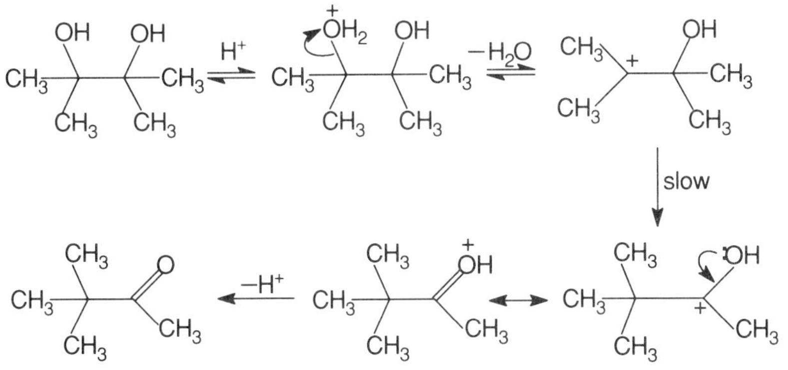

Scheme 4.1

The pinacol rearrangement can be used for the generation of spirocyclic systems (eqn 4.1).

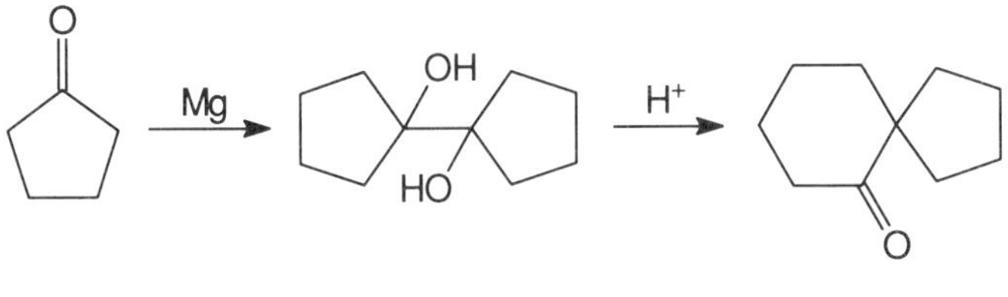

(4.1)

4.2 Acetal formation by 1,2- and 1,3-diols

Diols react with aldehydes and ketones under anhydrous conditions to give acetals and ketals (eqn 4.2). The reaction is acid-catalysed and proceeds most readily when the heterocyclic ring formed is five- or six-membered. Most examples therefore involve 1,2- or 1,3-diols. This transformation can be used to protect aldehydes and ketones, and diols, while other reactions involving base treatment or reduction are carried out elsewhere in the molecule (see chapter 10).

$$(4.2)$$

4.3 Dehydration of 1,4- and 1,5-diols

Diols can be cyclised to form cyclic ethers. This reaction can be carried out under acidic conditions and once again proceeds most readily when five- or six-membered ring formation is involved (eqn 4.3).

$$(4.3)$$

4.4 Oxidative cleavage of 1,2-diols

1,2-Diols are cleaved by periodate to give a dicarbonyl compound (eqn 4.4). The reaction involves a cyclic intermediate and therefore proceeds most easily when the two hydroxyl groups can adopt a *cis* or coplanar orientation (eqn 4.5).

$$(4.4)$$

$$(4.5)$$

Lead(IV) acetate can also be used to bring about oxidative cleavage of 1,2-diols. Once again a cyclic mechanism is implicated (eqn 4.6) so that a *cis* or coplanar arrangement is preferred (eqn 4.7).

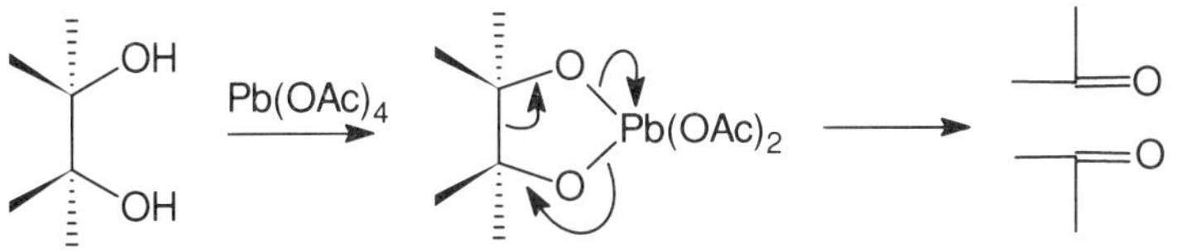

(4.6)

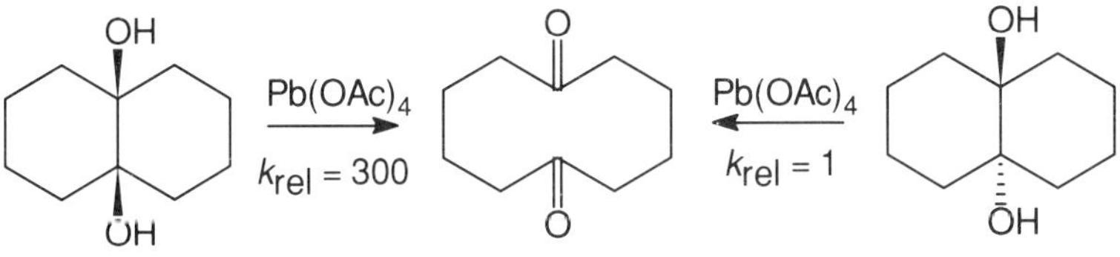

(4.7)

Problems

Suggest structures for compounds **A-H** in the following reaction sequences and give mechanisms for the reactions involved.

1.

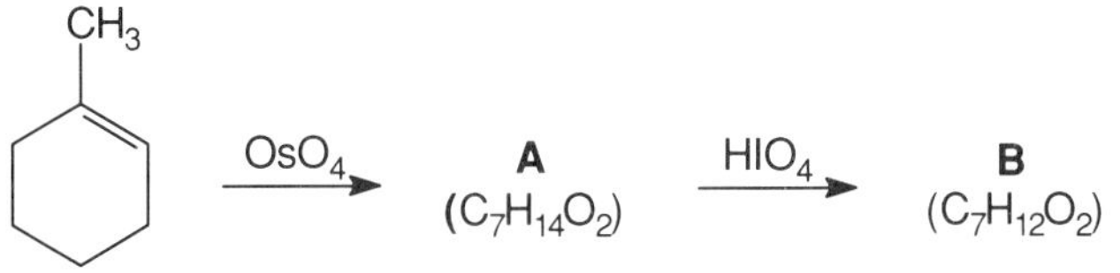

$$\xrightarrow{\text{OsO}_4} \quad \underset{(C_7H_{14}O_2)}{\textbf{A}} \quad \xrightarrow{\text{HIO}_4} \quad \underset{(C_7H_{12}O_2)}{\textbf{B}}$$

2.

$$\xrightarrow[\text{2. H}_3\text{O}^+]{\text{1. Mg}} \quad \underset{(C_{10}H_{22}O_2)}{\textbf{C}} \quad \xrightarrow{\text{H}^+} \quad \underset{(C_{10}H_{20}O)}{\textbf{D}}$$

3.

$$\xrightarrow[\text{2. }^-\text{OH/H}_2\text{O}]{\text{1. HCO}_3\text{H}} \quad \underset{(C_4H_{10}O_2)}{\textbf{E}} \quad \xrightarrow{\text{H}^+} \quad \underset{(C_9H_{16}O_2)}{\textbf{F}}$$

4.

$$\underset{(C_{10}H_{16})}{\textbf{G}} \quad \xrightarrow{\text{OsO}_4} \quad \underset{(C_{10}H_{18}O_2)}{\textbf{H}} \quad \xrightarrow{\text{HIO}_4} \quad$$

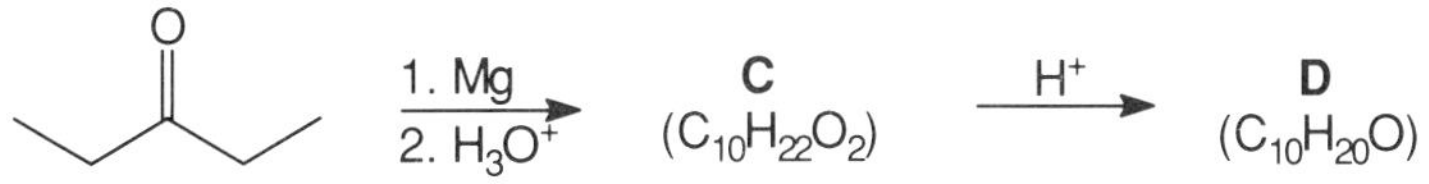

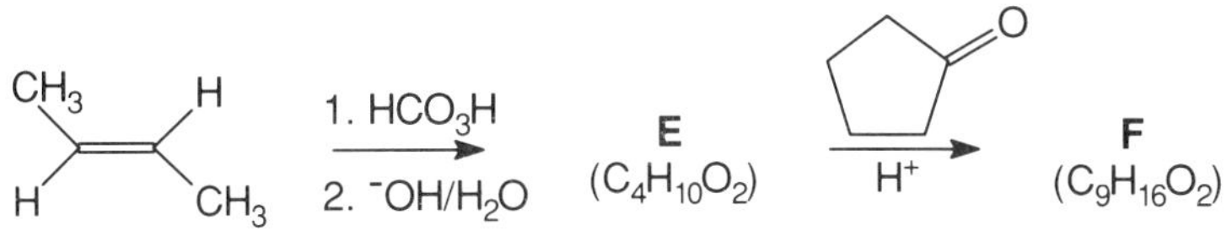

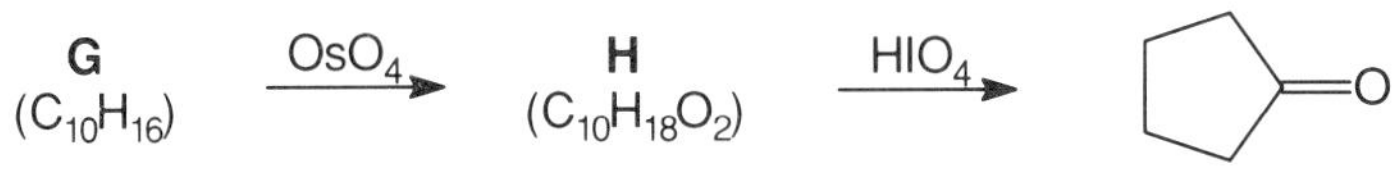

5 Reactions of hydroxy- and aminocarbonyl compounds

5.1 Hydroxyaldehydes and ketones

Hydroxyaldehydes and ketones react intramolecularly to form cyclic hemiacetals and, depending upon the size of the ring produced, this may, in some cases, be the thermodynamically preferred structure (e.g. eqn 5.1). When five- or six-membered rings are produced the cyclisation is so facile that it occurs even under neutral conditions.

$$HO(CH_2)_nCHO \rightleftharpoons \text{(cyclic hemiacetal)} \tag{5.1}$$

	% hemiacetal
$n = 3$	89
4	94
5	15

Since the open chain hydroxycarbonyl compound and the cyclic hemiacetal are in equilibrium in solution, compounds which formally do not display a carbonyl group may undergo the reactions of a carbonyl compound (scheme 5.1).

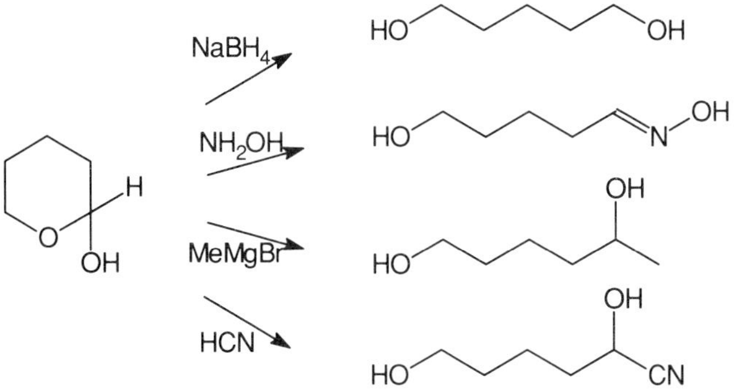

Scheme 5.1

Glucose is an example of a hydroxyaldehyde which exists mainly in the form of its cyclic hemiacetal, both in the solid state and in solution (eqns 5.2 and 5.3). The open chain aldehyde and the six-membered cyclic hemiacetal are in equilibrium in solution, so that glucose can react as an aldehyde, even though it exists mainly in the cyclic form. Indeed the fact that glucose contains an aldehyde group is reflected in its reducing properties (see later). The two diastereoisomeric cyclic hemiacetal forms of glucose are called *anomers*.

(5.2)

(5.3)

α-D-glucose
$[\alpha]_D = 112$

β-D-glucose
$[\alpha]_D = 19$

Furthermore two crystalline forms of D-glucose are known. The α form can be obtained by crystallisation from ethanol and the β form by crystallisation from pyridine. When α-D-glucose is dissolved in water the specific rotation of the solution falls gradually to an equilibrium value of 52.5. When β-D-glucose is dissolved in water the specific rotation rises to reach the same value. This phenomenon is called *mutarotation*. The proportions of the α and β anomers present at equilibrium are in the ratio 37:63.

Acetal formation

Just as aldehydes and ketones react with alcohols under anhydrous conditions to form acetals, so hydroxyaldehydes and ketones which form cyclic hemiacetals react with alcohols to form cyclic acetals. For example, on reaction with methanol and HCl glucose forms two acetals called *glucosides* (general name *glycosides*) (eqn 5.4).

D-glucose $\xrightarrow[\text{HCl}]{\text{MeOH}}$

(5.4)

methyl α-D-glucoside

methyl β-D-glucoside

Since glycosides are acetals rather than hemiacetals they are not in equilibrium with the aldehyde, they do not show reducing properties, and they do not exhibit mutarotation. However, in common with all acetals they are readily hydrolysed by aqueous acid, regenerating the corresponding monosaccharide.

Oxidation

α-Hydroxyaldehydes can be readily oxidised, and many α-hydroxyketones also give a positive test with Fehling's solution and Tollen's reagent since, in alkaline solution, they are in equilibrium with the isomeric α-hydroxy-aldehyde (eqn 5.5). Thus in the carbohydrate series both *aldoses* (e.g. glucose) and *ketoses* (e.g. fructose) respond to this test. Furthermore, glucose is easily oxidised to the corresponding lactone (eqn 5.6).

(5.5)

(5.6)

α-Hydroxyaldehydes and ketones are also susceptible to periodate oxidation (eqns 5.7–5.11), and this procedure was instrumental in much of the early work by Fischer and others in determining the structures of these highly functionalised compounds.

For reaction of periodate with 1,2-diols, see p.39.

$$\underset{R}{\overset{OH}{\underset{O}{\bigvee}}}\!\!R' \xrightarrow{HIO_4} RCHO \ + \ R'CO_2H \tag{5.7}$$

$$\text{(eqn)} \xrightarrow{HIO_4} RCHO \ + \ HCO_2H \ + \ R'CHO \tag{5.8}$$

$$\text{(eqn)} \xrightarrow{HIO_4} RCHO \ + \ CO_2 \ + \ R'CHO \tag{5.9}$$

$$\text{D-glucose} \xrightarrow{HIO_4} 5\,HCO_2H \ + \ CH_2O \tag{5.10}$$

$$\text{D-fructose} \xrightarrow{HIO_4} 3\,HCO_2H \ + \ 2\,CH_2O \ + \ CO_2 \tag{5.11}$$

Osazone formation

α-Hydroxyaldehydes and ketones react with phenylhydrazine to give a bis-phenylhydrazone called an osazone (eqn 5.12). Osazone formation consumes three equivalents of phenylhydrazine, one of which is reduced to aniline (eqn 5.13).

$$
\begin{array}{c}
\text{CHO} \\
\text{—OH} \\
\text{R}
\end{array}
\xrightarrow{3\text{PhNHNH}_2}
\begin{array}{c}
\text{CH=NNHPh} \\
\text{=NNHPh} \\
\text{R}
\end{array}
+ \quad \text{PhNH}_2
\qquad (5.12)
$$

$$
\begin{array}{c}
\text{CH=NNHPh} \\
\text{H—} \quad \text{—OH} \\
\text{R}
\end{array}
\longrightarrow
\begin{array}{c}
\text{CH—NH—NHPh} \\
\text{O} \\
\text{R}
\end{array}
\xrightarrow[-\text{PhNH}_2]{}
\begin{array}{c}
\text{CH=NH} \\
\text{—O} \\
\text{R}
\end{array}
\xrightarrow{2\text{PhNHNH}_2}
\begin{array}{c}
\text{CH=NNHPh} \\
\text{=NNHPh} \\
\text{R}
\end{array}
\qquad (5.13)
$$

Retro-aldol reactions and dehydration

Since the aldol reaction is reversible β-hydroxyaldehydes and ketones will in some cases undergo retro-aldol reactions on treatment with base (eqn 5.14). Dehydration of β-hydroxycarbonyl compounds also occurs very easily because the proton removed is relatively acidic and the product contains a conjugated π system (see also section 2.3).

$$ (5.14) $$

E1cB mechanism under basic conditions, see p.20.

5.2 Hydroxyacids

Just as hydroxyaldehydes cyclise to form hemiacetals, hydroxyacids can be cyclised to give lactones (eqn 5.15). Indeed, γ-hydroxyacids usually spontaneously cyclise to give five-membered γ-lactones (eqns 5.16–5.17). Lactones of other ring sizes can also be prepared from the corresponding hydroxyacids if the water is removed as it is formed.

$$
\text{HO(CH}_2)_n\text{CO}_2\text{H} \; \underset{}{\overset{\text{H}^+}{\rightleftharpoons}} \;
\begin{array}{c}
(\text{CH}_2)_{n-2} \\
\end{array}
+ \quad \text{H}_2\text{O}
\qquad (5.15)
$$

% lactone

n	
2	0
3	73
4	9
5	0

(5.16)

(5.17)

α-Hydroxyacids react to give a six-membered lactone rather than the alternative three-membered ring compound (eqn 5.18). Dehydration of β-hydroxyacids ($n = 2$ in eqn 5.15) normally leads to the αβ-unsaturated acid rather than the four-membered β-lactone.

(5.18)

5.3   Amino acids

α-Amino acids exist almost exclusively as zwitterions and as a result they are high melting solids which are usually water soluble. In acidic solution they form cations which move towards the cathode in an electric field, whereas in alkaline solution they form anions which move towards the anode (eqn 5.19). This process forms the basis of *electrophoresis* which is a technique used for analysing and separating amino acids. The pH at which a solution of an amino acid is electrically neutral varies from compound to compound. This pH, at which there are equal numbers of cations and anions, is called its *isoelectric point*.

(5.19)

α-Amino acids form esters on treatment with methanol and HCl. On storage or on heating the esters derived from α-amino acids are converted into six membered bis-lactams, called diketopiperazines (eqn 5.20).

(5.20)

α-Amino acids can also be acylated and the derivatives obtained cyclise to form azlactones (eqn 5.21).

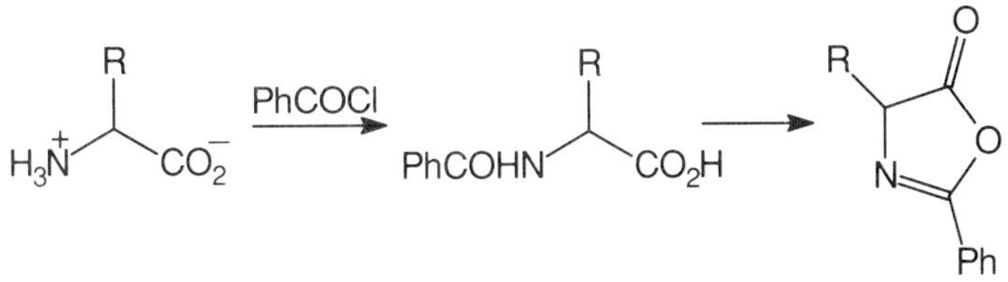

(5.21)

A very useful property of α-amino acids is their ability to form a purple coloured derivative with ninhydrin (eqn 5.22), which is therefore used as a spray reagent to locate spots due to α-amino acids on paper chromatography.

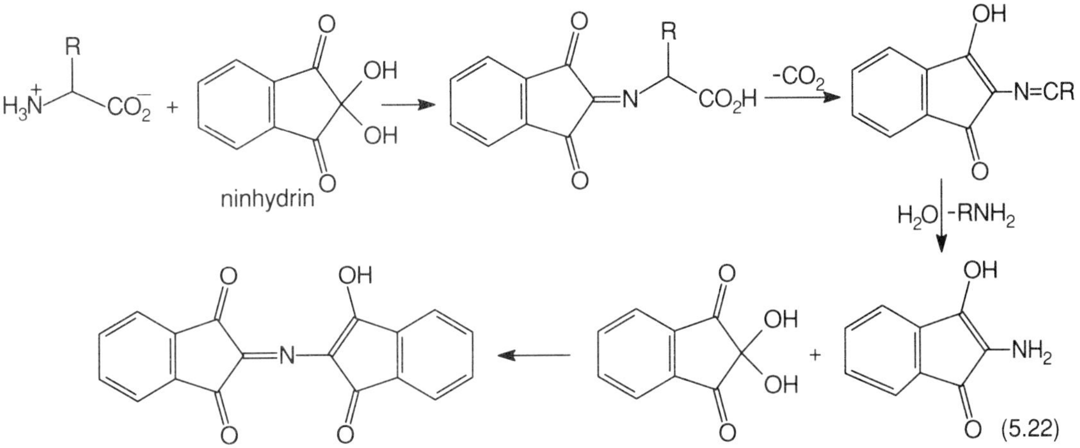

(5.22)

Problems

1. Suggest steps to carry out the following transformation.

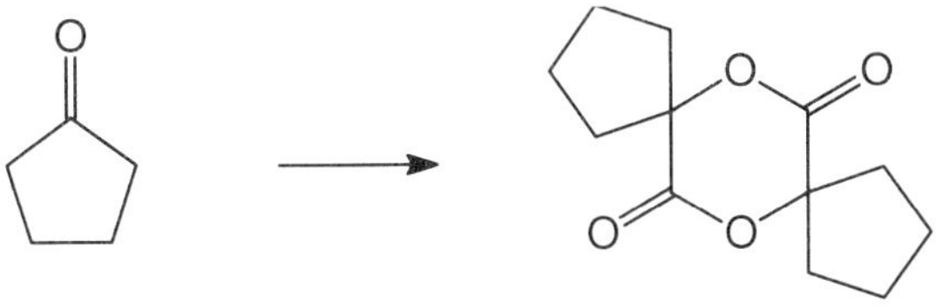

2. Predict the product of the following reaction.

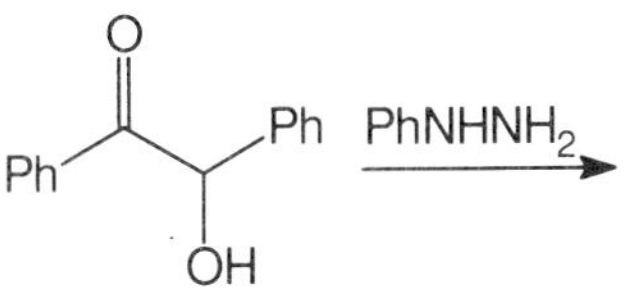

3. Suggest a mechanism for the following reaction.

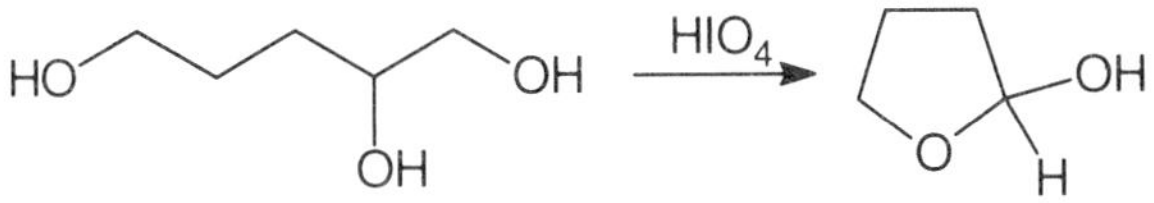

6 Reactions of dicarbonyl compounds

6.1 Reactions of α-dicarbonyl compounds

Benzilic acid rearrangement

α-Dicarbonyl compounds undergo rearrangement on treatment with alkali to give α-hydroxyacids (eqn 6.1). The reaction provides a useful synthetic route to some α-hydroxyacids. The mechanism involves addition of hydroxide to one of the carbonyl groups, followed by migration of one of the alkyl or aryl substituents to the second carbonyl group (eqn 6.2) (see L. M. Harwood, *Polar rearrangements;* in this series, for further details).

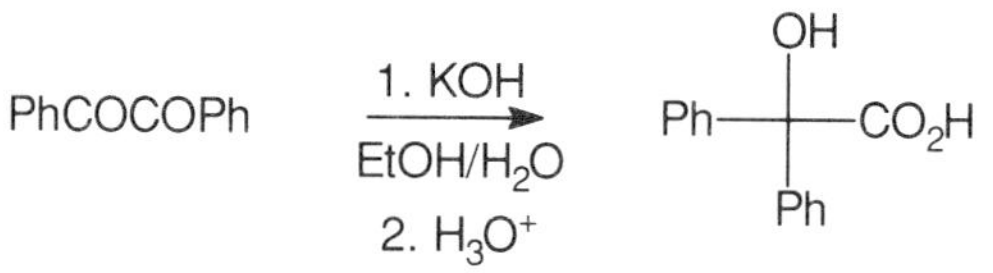

$$
\text{PhCOCOPh} \quad \xrightarrow[\substack{\text{EtOH/H}_2\text{O} \\ \text{2. H}_3\text{O}^+}]{\text{1. KOH}} \quad \underset{\text{Ph}}{\overset{\text{OH}}{\text{Ph}{-}\!\!{-}\text{CO}_2\text{H}}} \tag{6.1}
$$

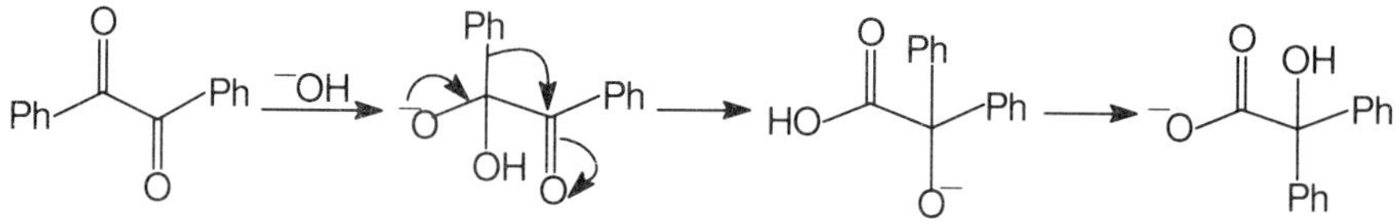

$$\tag{6.2}$$

Enolisation

It is of interest to contrast the structure of 2,3-butanedione, which exists almost entirely in the diketo form, with that of 1,2-cyclopentanedione, which exists almost entirely as the enol. A suggested explanation for this difference is that 2,3-butanedione can adopt an *anti* conformation with the two carbonyl dipoles diametrically opposed, while in the diketo form 1,2-cyclopentanedione is forced to adopt a structure in which there is a strong repulsive interaction between the two carbonyl groups. This arrangement is avoided in the monoenol tautomer (scheme 6.1).

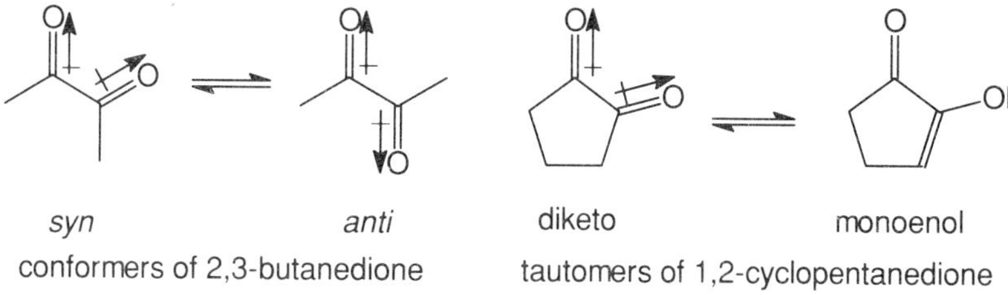

| *syn* | *anti* | diketo | monoenol |

conformers of 2,3-butanedione   tautomers of 1,2-cyclopentanedione

Scheme 6.1

6.2   Reactions of β-dicarbonyl compounds

Enolisation

β-Dicarbonyl compounds exist in equilibrium with their enol tautomers and in some solvents, which are themselves incapable of hydrogen bonding, the enol tautomer may be the more stable and therefore the major form present (eqns 6.3 and 6.4).

(6.3)

in water	84	:	16
in hexane	8	:	92

(6.4)

in water	90	:	10
in hexane	51	:	49

Acidity

β-Dicarbonyl compounds are more acidic than the analogous monofunctional compounds (scheme 6.2). This is explained by the greater stability of the corresponding conjugate base due to resonance stabilisation by the two carbonyl groups.

$$CH_3COCH_2COCH_3 \qquad CH_3COCH_2CO_2Et \qquad CH_2(CO_2Et)_2$$
$$pK_a \ 9 \qquad\qquad pK_a \ 11 \qquad\qquad pK_a \ 13$$

$$cf. \quad CH_3COCH_3 \qquad\qquad CH_3CO_2Et$$
$$pK_a \ 20 \qquad\qquad\qquad pK_a \ 24$$

Scheme 6.2

Alkylation

The acidity of β-dicarbonyl compounds can be used to advantage in synthesis since the enolate anions produced can be generated and alkylated under relatively mild conditions (eqns 6.5–6.8).

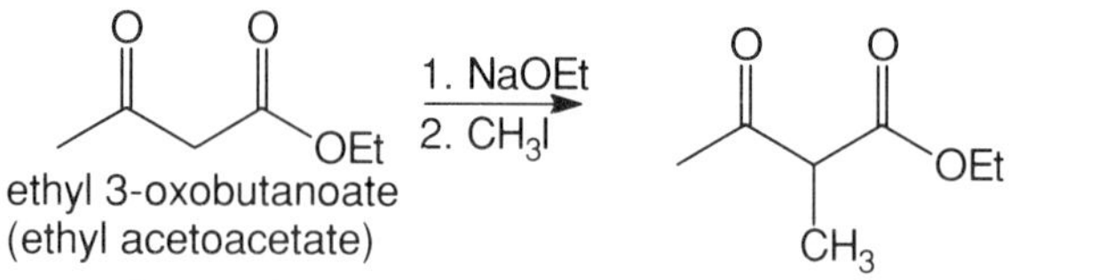

ethyl 3-oxobutanoate
(ethyl acetoacetate)

1. NaOEt
2. CH₃I

(6.5)

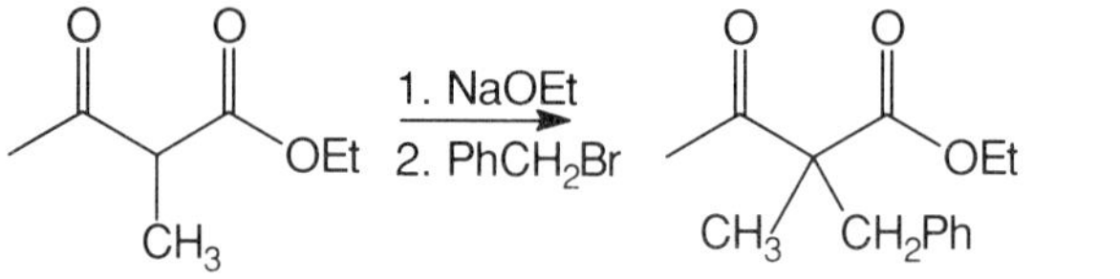

CH₃

1. NaOEt
2. PhCH₂Br

CH₃   CH₂Ph

(6.6)

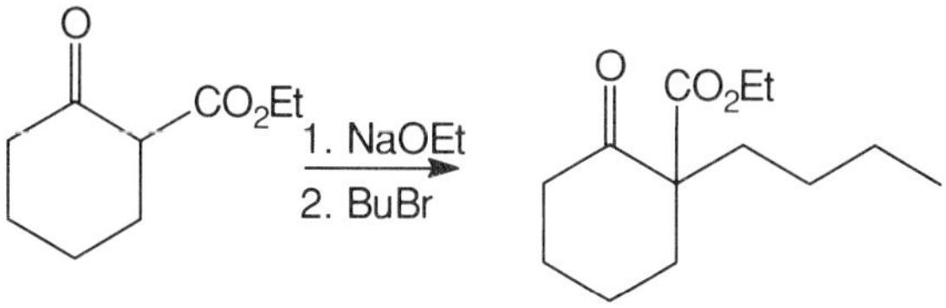

CH₃

CO₂Et
1. NaOEt
2. BuBr

CO₂Et

(6.7)

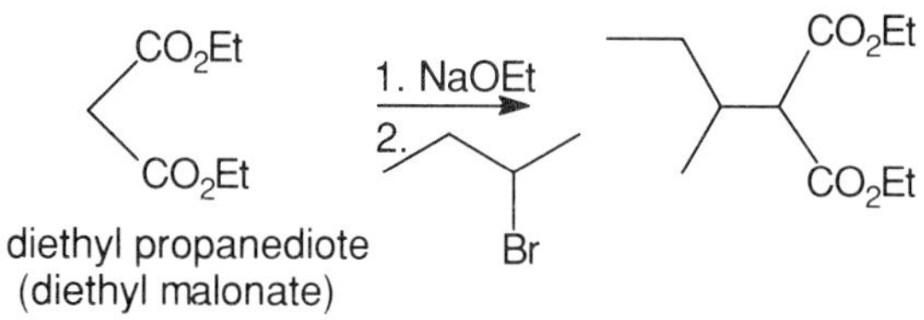

CO₂Et
1. NaOEt
2.

CO₂Et

CO₂Et
CO₂Et

(6.8)

diethyl propanediote
(diethyl malonate)

Br

Decarboxylation

β-Ketoacids and propanedioic acids which can be readily obtained by hydrolysis of the corresponding esters undergo decarboxylation on heating to give ketones and carboxylic acids respectively (eqns 6.9 and 6.10).

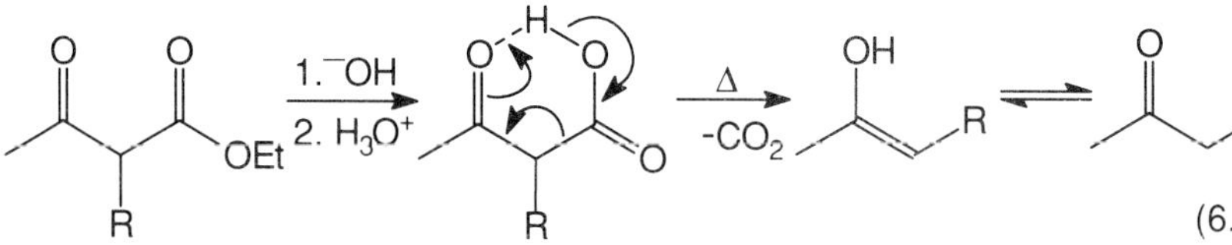

1. ⁻OH
2. H₃O⁺

Δ
-CO₂

R   R

(6.9)

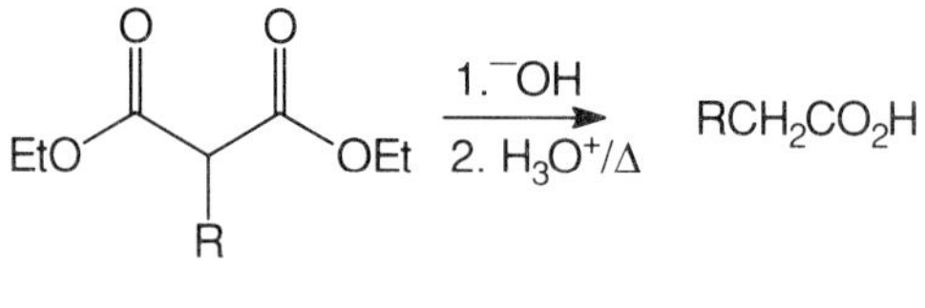

OEt

R   R

OH

1. ⁻OH
2. H₃O⁺/Δ

RCH₂CO₂H

(6.10)

EtO   OEt

R

Use of β-ketoesters and related compounds in synthesis

By combining the above properties of β-dicarbonyl compounds it is possible to devise a protocol which is very useful for the synthesis of substituted carbonyl compounds. For example, ethyl 3-oxobutanoate (ethyl acetoacetate) and diethyl propanedioate (diethyl malonate) can be regarded as *synthons* for the synthesis of methyl ketones and substituted ethanoic acid derivatives respectively (scheme 6.3).

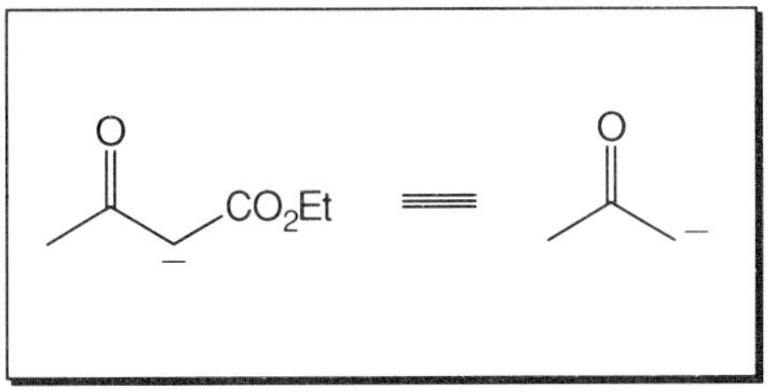

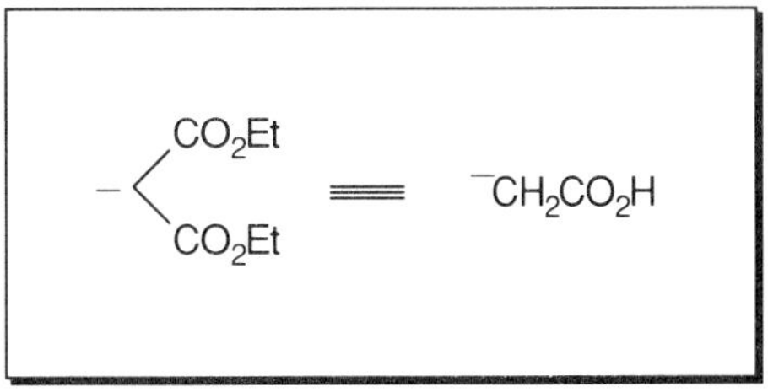

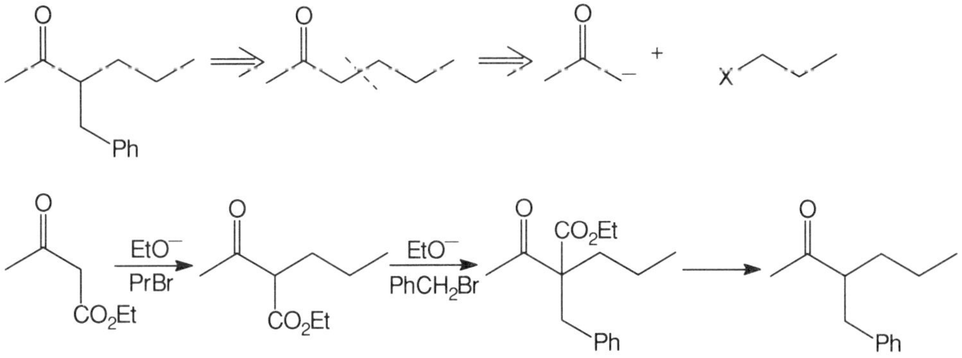

Scheme 6.3

Ring synthesis

By employing a dihaloalkane as the alkylating agent it is possible to use ethyl acetoacetate or diethyl malonate to synthesise alicyclic compounds (e.g. eqn 6.11).

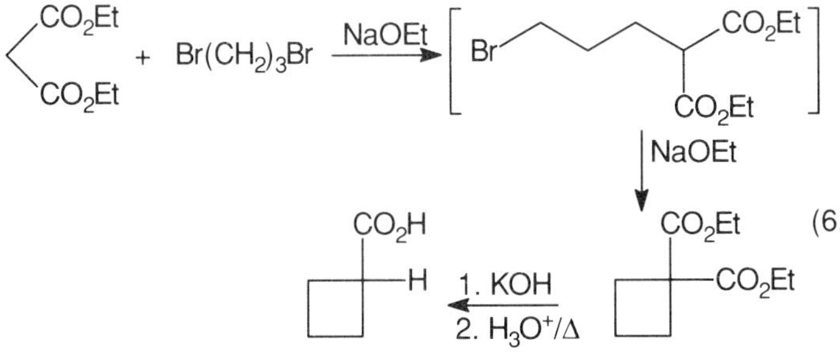

$$(6.11)$$

Amino acid synthesis

Bromination of diethyl malonate followed by reaction with potassium phthalimide provides a useful method for the synthesis of α-amino acids (eqn 6.12).

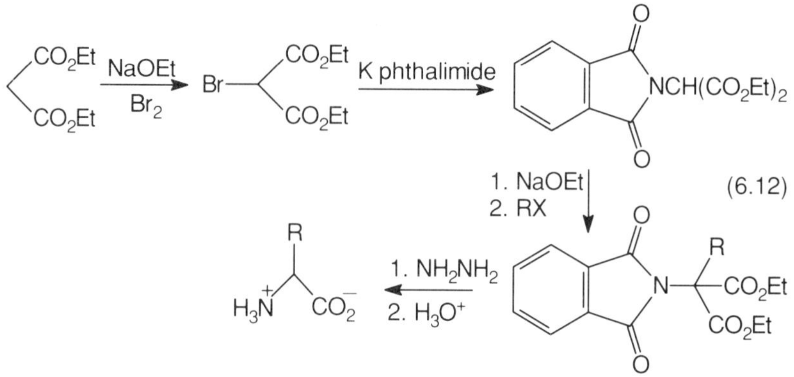

(6.12)

For mechanism of reaction with
hydrazine, see p.13.

A second method for α-amino acid synthesis using diethyl malonate involves nitrosation followed by reduction and acylation to give an acylated aminomalonate which can then be alkylated (eqn 6.13).

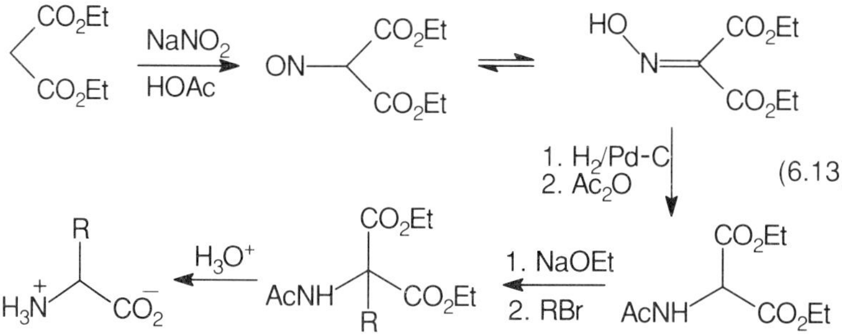

(6.13)

Synthesis of 1,4-dicarbonyl compounds

By coupling together two molecules of a β-dicarbonyl compound it is possible to obtain a 1,4-dicarbonyl compound. This can be achieved by halogenation of the β-dicarbonyl compound followed by reaction with a second molecule of the β-dicarbonyl compound (eqn 6.14).

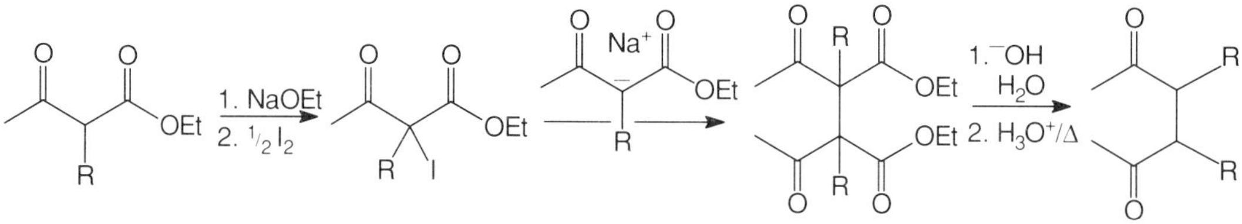

(6.14)

Alternatively, reaction of a β-ketoester with an α-halocarbonyl compound, followed by hydrolysis and decarboxylation provides a route to unsymmetrically substituted 1,4-dicarbonyl compounds (eqn 6.15).

(6.15)

Knoevenagel reaction

The Knoevenagel reaction involves the condensation of an aldehyde with diethyl malonate, in the presence of base, to give an αβ-unsaturated carbonyl compound (eqn 6.16). When followed by ester hydrolysis and decarboxylation it provides an alternative to the aldol condensation and the Reformatsky reaction for the synthesis of αβ-unsaturated acids. However, other compounds containing acidic hydrogen atoms can also be used as substrates so that it in fact typifies a much more general procedure (e.g. eqn 6.17).

$$PhCHO \ + \ CH_2(CO_2Et)_2 \ \xrightarrow{\text{piperidine}} \quad (6.16)$$

piperidine

(6.17)

Alkylation of dianions of ethyl acetoacetate

Although most of the well-known reactions of β-ketoesters involve reaction at the α position, it is possible to bring about selective reaction at the γ position by abstracting a second proton using a stronger base. Since the γ anion is less stable than the α anion, reaction takes place first at the more reactive γ position (eqn 6.18).

(6.18)

Retro-Claisen reaction

It should be recalled that β-dicarbonyl compounds can undergo a retro-Claisen reaction on treatment with base (eqn 6.19). This is most likely to occur when the dicarbonyl compound has no α-hydrogen atoms, removal of which would give a stabilised carbanion, and when the base is a good nucleophile. The retro-Claisen reaction involves nucleophilic addition to one of the carbonyl groups, followed by carbon–carbon bond cleavage.

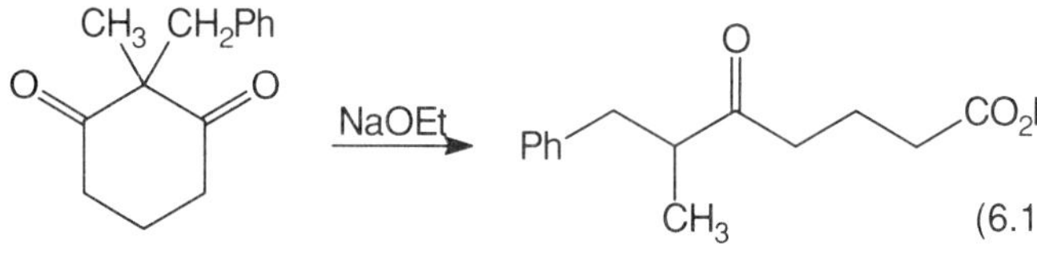

(6.19)

6.3   Reactions of 1,4-dicarbonyl compounds

Two useful reactions of 1,4-dicarbonyl compounds are the cyclisation of the diketones to give furan, pyrrole and thiophene derivatives, and the Stobbe condensation of the diesters with aromatic aldehydes to give unsaturated dicarboxylic acid derivatives (eqns 6.20–6.23).

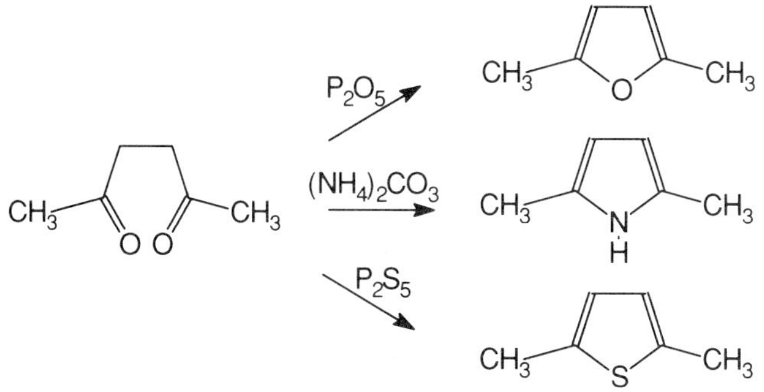

(6.20)

(6.21)

(6.22)

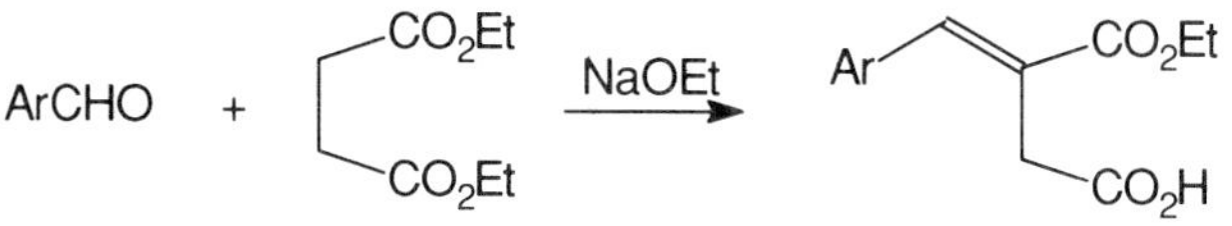

(6.23)

The outcome of the Stobbe condensation is unusual in that it yields the monoester of the butanedioic acid. The mechanism proposed for this reaction must clearly account for this observation and indeed the simplest explanation is that depicted in scheme 6.4 in which cyclisation to give an intermediate lactone precedes the elimination to give the double bond.

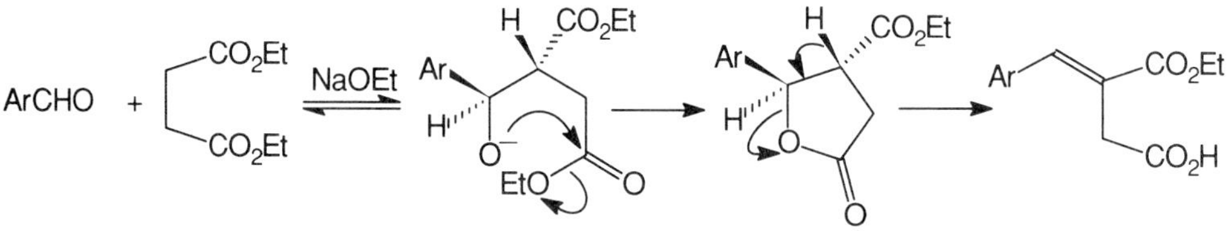

Scheme 6.4

Problems

1. Predict the products of each of the following reactions.

(a)

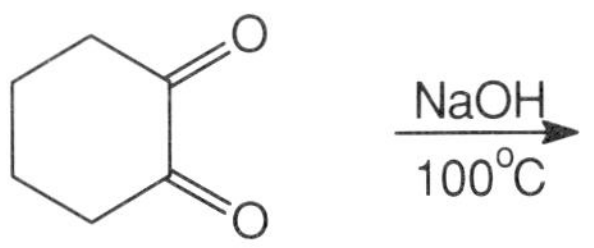

$\xrightarrow[100°C]{NaOH}$

(b)

CHO
|
CHO

$\xrightarrow[100°C]{NaOH}$

2. Suggest methods by which each of the following compounds may be synthesised from the suggested starting material.

(a)

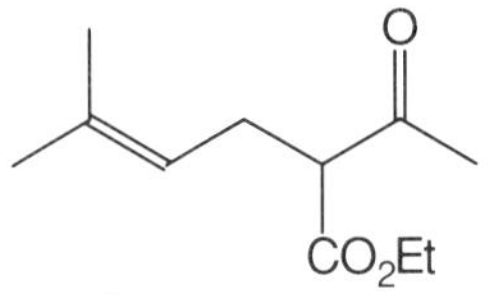

CO₂Et

from

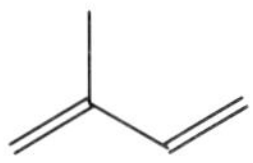

(b)

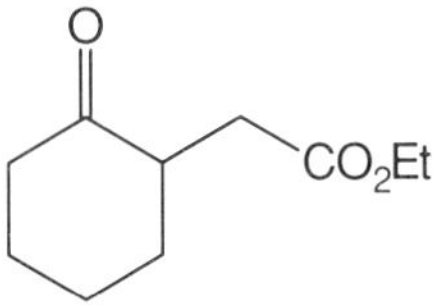

CO₂Et

from

CO₂Et

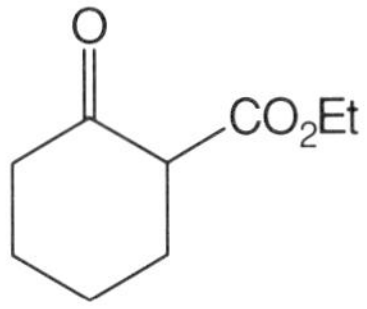

7 Reactions of unsaturated carbonyl compounds

7.1 Reactions of ketenes

Ketene itself is stable in the gas phase but dimerises on liquifaction (eqn 7.1).

$$\text{(7.1)}$$

β-lactone

Ketenes undergo two characteristic reactions, which are also typical of so-called heterocumulenes. These are nucleophilic addition (eqn 7.2) and cycloaddition (eqn 7.3). Reactions with nucleophiles result in acetylation (or more generally acylation) of the nucleophilic reagent. The cycloaddition reactions of ketenes are similar to the Diels–Alder reaction (see section 3.3), although they are classified as [2+2] cycloaddition reactions, in contrast to the Diels–Alder reaction which is a [4+2] process. In the example shown (eqn 7.3), reaction with an alkoxyalkyne gives a cyclobutenone. Since this is an enol ether (see section 8.2) it is readily hydrolysed, giving a β-diketone. The dimerisation of ketene (eqn 7.1) is a very similar reaction, although in this case one ketene molecule undergoes cycloaddition across the carbon–oxygen double bond.

$$CH_2={=}O \longrightarrow CH_3-C\overset{O}{\underset{Y}{\diagup}} \qquad \text{(7.2)}$$

$$H\ddot{Y} \qquad (HY = H_2O,\ NH_3,\ ROH,\ RNH_2)$$

$$\xrightarrow{\Delta} \xrightarrow{H_3O^+} \qquad \text{(7.3)}$$

β-diketone

7.2 Reactions of αβ-unsaturated carbonyl compounds

The chemistry of αβ-unsaturated carbonyl compounds, like that of conjugated dienes, is dependent upon the interaction between the orbitals of the carbon–carbon double bond and those of the carbon–oxygen double bond. In particular the electron-withdrawing carbonyl group has a profound influence on the reactivity of the carbon–carbon double bond.

Reaction with electrophiles

The conjugated carbonyl group *deactivates* the carbon–carbon double bond towards *electrophiles*. αβ-Unsaturated carbonyl compounds are therefore *less reactive* than simple alkenes towards electrophilic reagents. They do nevertheless undergo addition reactions with electrophiles, but less readily than simple alkenes (eqns 7.4 and 7.5).

$$(7.4)$$

$$(7.5)$$

Reaction with nucleophiles

In contrast, the conjugated carbonyl group *activates* the carbon–carbon double bond towards *nucleophiles*. αβ-Unsaturated carbonyl compounds are therefore *more reactive* than simple alkenes towards nucleophilic reagents. Indeed, in contrast to simple alkenes, they undergo nucleophilic addition reactions (eqns 7.6–7.8).

$$(7.6)$$

$$(7.7)$$

$$(7.8)$$

This process is described as 1,4-addition or *conjugate addition* to distinguish it from the usual 1,2-addition reaction of aldehydes and ketones. Indeed, 1,2- and 1,4-addition reactions compete with one another and reactions of αβ-unsaturated carbonyl compounds usually give mixtures of 1,2- and 1,4-addition products. An intriguing aspect of the chemistry of these compounds is the competition between these two modes of reaction and the factors which control their relative importance.

Since, as we will see, the competition between the two modes of reaction is influenced by several factors including the reagent, reaction mechanism, and reaction conditions, formulating clear guidelines for predicting the outcome of such reactions is difficult. One rule of thumb which can be used in many cases, is to classify nucleophiles as either hard or soft. Anions in which the negative charge is localised on a small atom are classified as *hard nucleophiles* (e.g. $^-NH_2$, RO^- and H^-). Such nucleophiles give mainly 1,2-addition because they attack the carbonyl group directly and irreversibly, since they are usually strong bases and therefore are poor leaving groups. Neutral nucleophiles (e.g. amines) and those in which the negative charge is delocalised (e.g. resonance-stabilised carbanions), or located on a large atom (e.g. RS^-), are classified as *soft nucleophiles*. They are usually weak bases and are therefore good leaving groups. As a result they add reversibly to the carbonyl group and lead ultimately to the thermodynamically more stable 1,4-addition product.

Scheme 7.1

Addition of organometallic reagents

Grignard reagents usually give a mixture of 1,2- and 1,4-addition products in which the proportions vary depending upon the steric constraints imposed (Scheme 7.2).

	1,2-product	1,4-product
PhCH=CHCHO	100%	0%
PhCH=CHCOCH$_3$	40%	60%
PhCH=CHCOPh	1%	99%
PhCH=C(Ph)COPh	0%	100%

Scheme 7.2

Organolithium reagents on the other hand give exclusively 1,2-addition, while organocopper reagents give mainly 1,4-addition. By careful choice of reagent it is therefore possible to achieve either 1,2- or 1,4-addition. As a result of their usefulness in bringing about 1,4-addition, the chemistry of organocopper reagents has been extensively developed, and several different types are available. Two of the simplest and most widely utilised are the lithium dialkylcuprates and alkyl copper species (scheme 7.3).

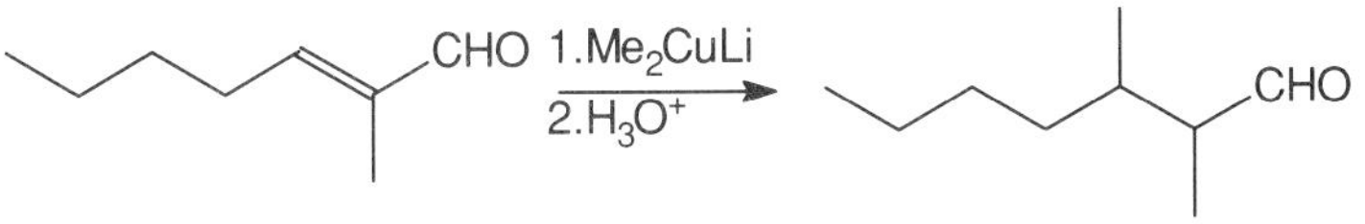

$$CH_3Li \; + \; CuI \longrightarrow CH_3Cu \; + \; LiI$$
$$2CH_3Li \; + \; CuI \longrightarrow (CH_3)_2CuLi \; + \; LiI$$

Scheme 7.3

The state of aggregation of the lithium organocuprates is uncertain, but it seems likely that, in many, the organic ligands are bonded to tetrahedral clusters of four metal atoms. Spectroscopic evidence suggests that in ether solution lithium dimethylcuprate exists as a dimer.

The reason for the high selectivity in favour of 1,4-addition by organocopper reagents is still a matter for debate, but the reaction mechanism is thought to involve an initial electron transfer from the organocopper species to the ketone followed by coupling and intramolecular transfer of the organic group from the metal to the β-carbon atom. Certainly there is good evidence for the involvement of an enolate anion since this can be intercepted not only by a proton but also by other electrophiles such as alkylating agents (scheme 7.4).

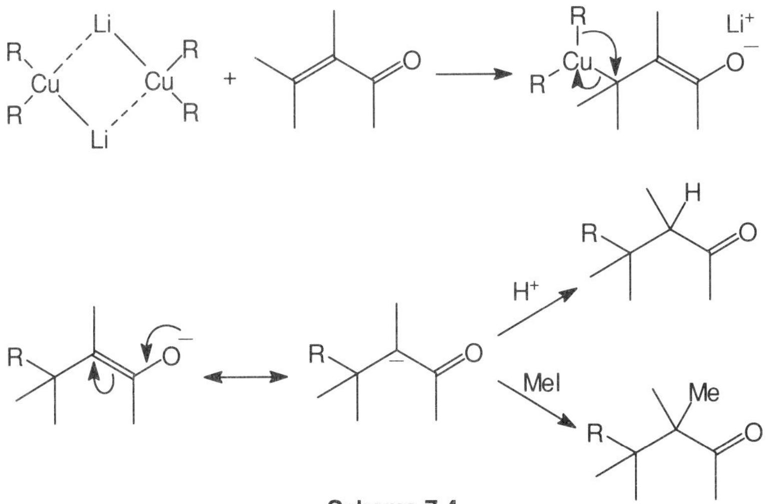

Scheme 7.4

One disadvantage of the use of simple lithium organocuprates $LiCuR_2$ is that only one of the two organic groups R is transferred. Very often this may not matter, but in cases where the alkyl group to be introduced has been arduously synthesised it is an advantage to use 'mixed' reagents in which one expendable group is tightly bound to copper and not transferred.

Higher yields are sometimes obtained by using so-called higher-order cuprates such as $Li_2R_2Cu(CN)$ obtained by reacting two equivalents of the organolithium compound with copper(I) cyanide, or by carrying out the reaction in the presence of a Lewis acid. Interestingly, addition of catalytic amounts of copper(I) salts also drives the reaction with Grignard reagents in favour of 1,4-addition (eqn 7.9).

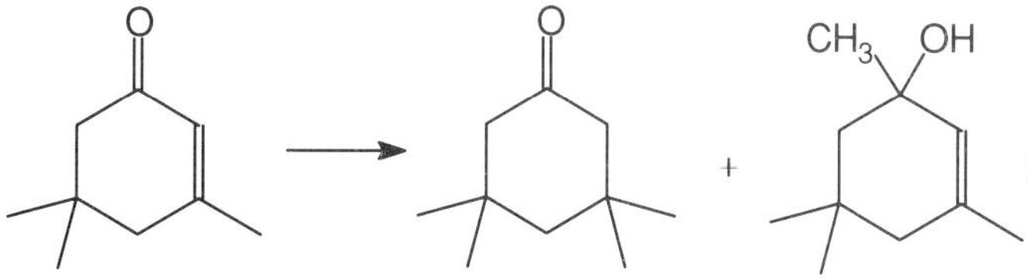

CH_3MgBr	1.5%	98.5%
CH_3MgBr + CuCl	82.5%	17.5%

Addition of carbanions–Michael addition

One of the most useful and well-known reactions of αβ-unsaturated carbonyl compounds is their reaction with resonance stabilised carbanions, called Michael addition. The mechanism of this reaction is shown in scheme 7.5. When the carbanion is derived from a carbonyl compound the product obtained is a 1,5-dicarbonyl compound (eqns 7.10 and 7.11). However a wide variety of carbanions can be employed (e.g. eqn 7.12).

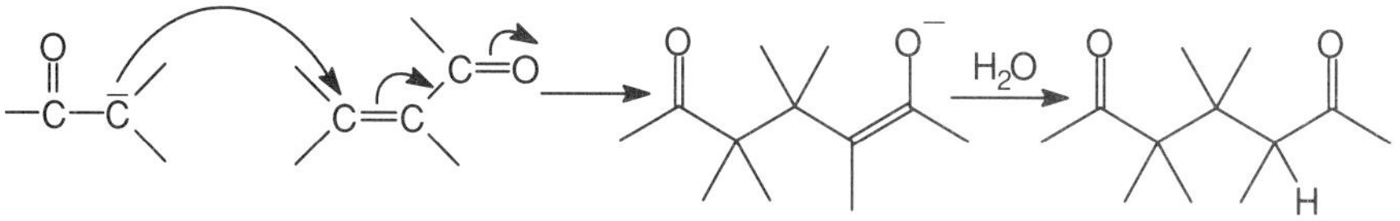

Scheme 7.5

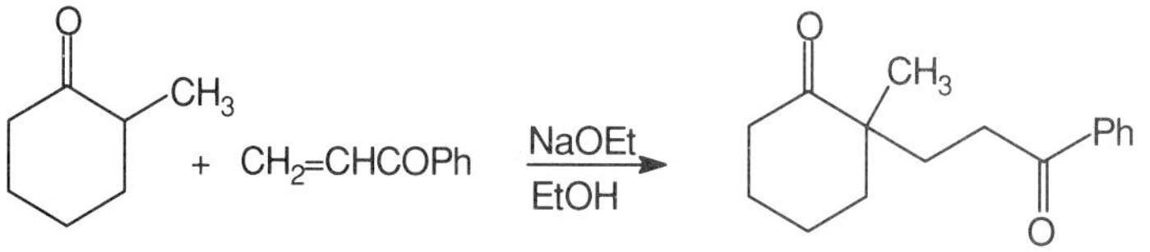

(7.10)

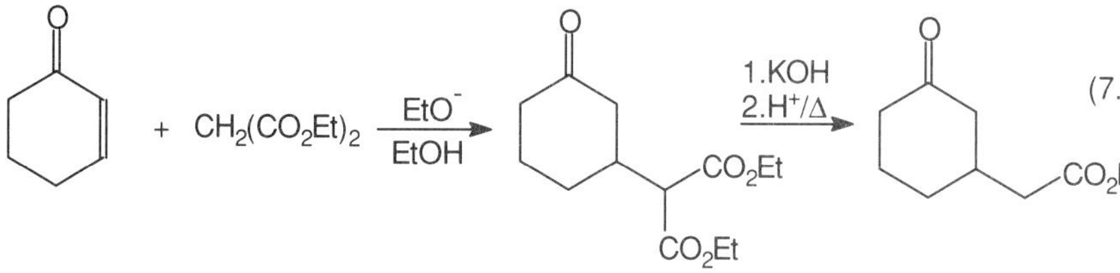

(7.11)

For conversion of nitro to carbonyl group, see p.10.

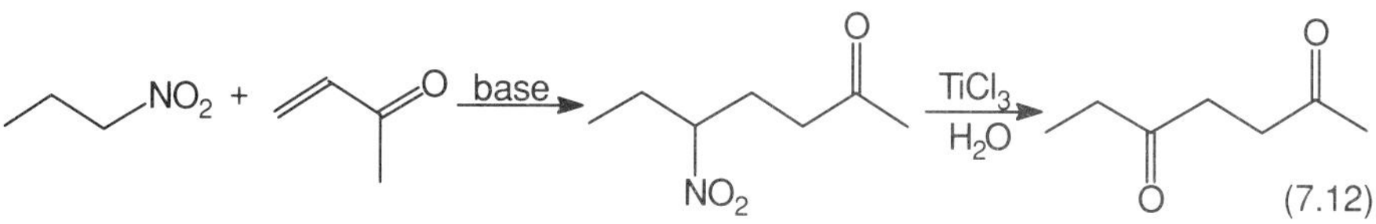

(7.12)

A special case arises when the anion is derived from a cyclohexanone and the unsaturated carbonyl compound is butenone (scheme 7.6). In this case a second base-catalysed reaction, an aldol condensation, can occur leading to the formation of a second six-membered ring. This two step process is called *Robinson annelation* and, along with the Diels–Alder reaction, is one of the most important methods for the preparation of structures, such as steroids, containing fused six-membered rings.

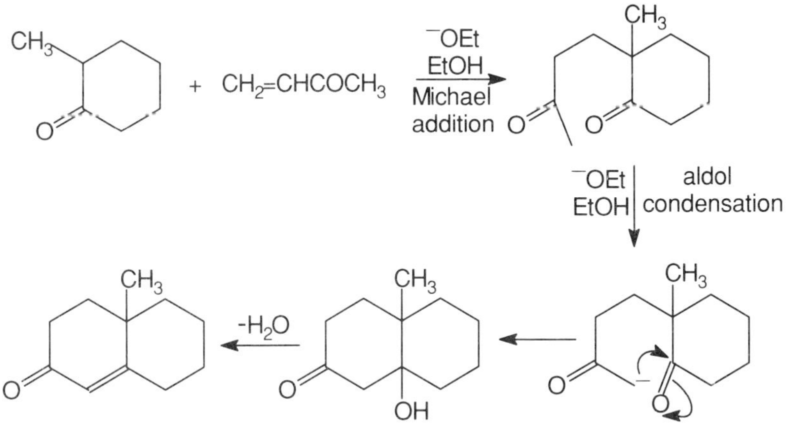

Scheme 7.6

Dissolving metal reduction

αβ-Unsaturated carbonyl compounds can be reduced using lithium in liquid ammonia (eqn 7.13). This reaction has the advantage that it is more selective than catalytic hydrogenation (e.g. eqn 7.14), and also provides yet another useful method for regioselectively generating enolates of carbonyl compounds, which can be trapped by reaction with electrophiles (e.g. eqns 7.15 and 7.16).

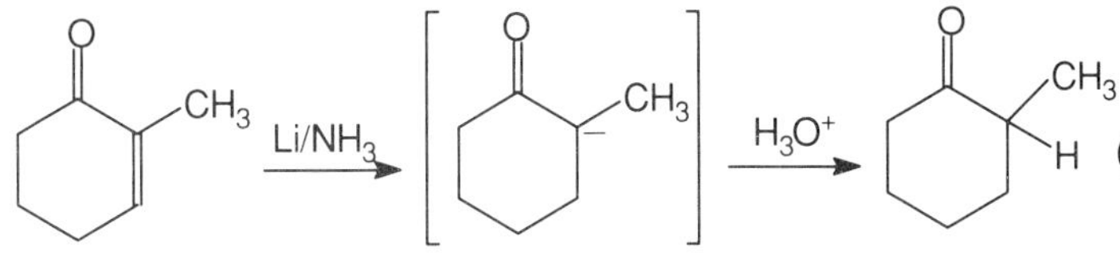

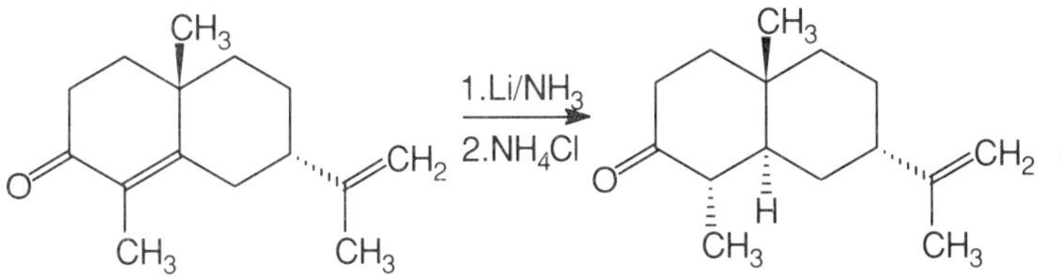

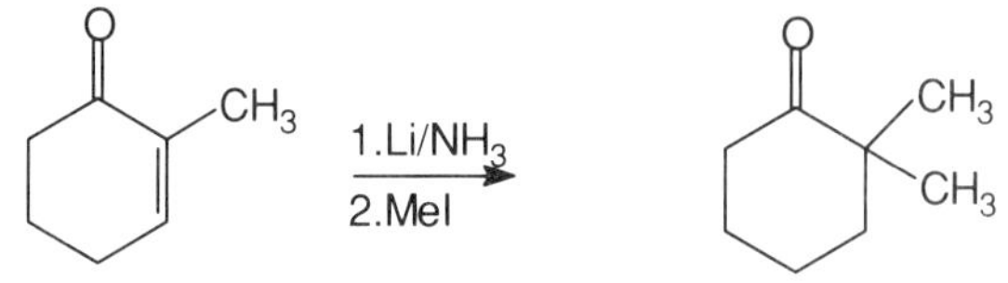

$$(7.15)$$

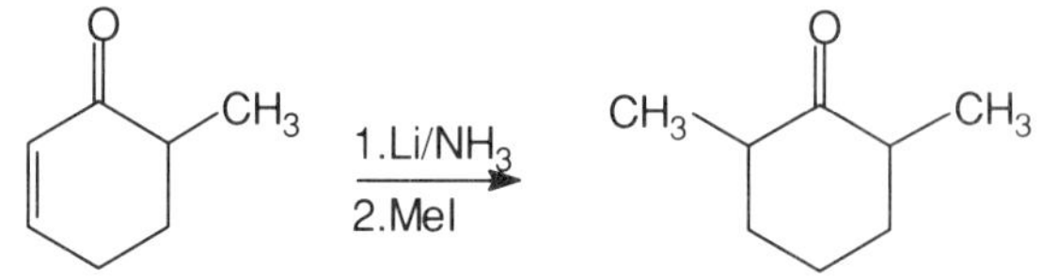

$$(7.16)$$

The mechanism of the reaction is analogous to that of the dissolving metal reduction of aromatic compounds and involves electron transfer from the metal followed by protonation by ammonia (scheme 7.7).

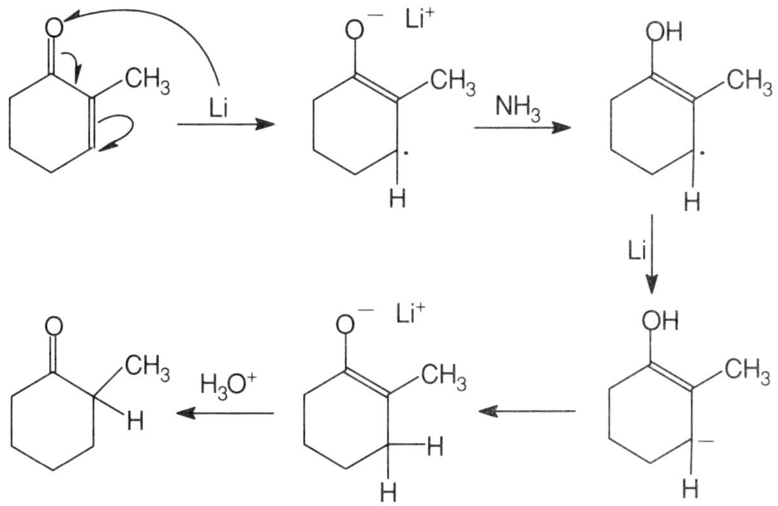

Scheme 7.7

Photochemical cycloaddition

αβ-Unsaturated ketones readily undergo photochemical cycloaddition reactions (e.g. eqn 7.17). These reactions, like the thermally induced cycloaddition reactions of ketenes, are of the [2+2] type. They are thought to proceed by attack of an $n \rightarrow \pi^*$ triplet on another ground state molecule.

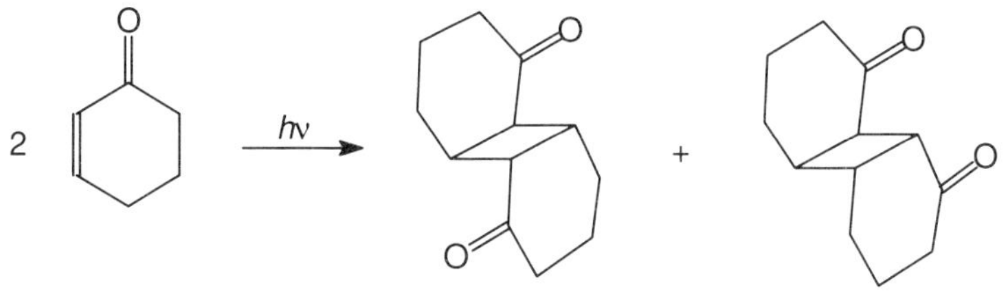

$$(7.17)$$

7.3 Reactions of quinones

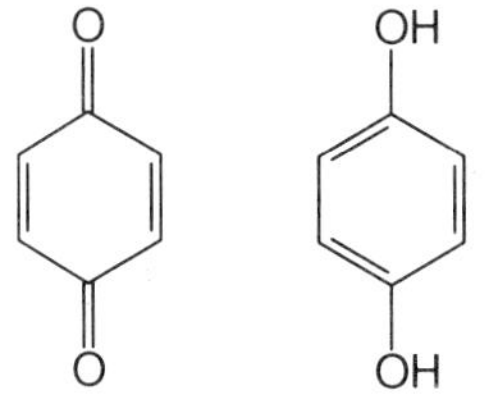

Quinones are examples of αβ-unsaturated carbonyl compounds and as such they undergo conjugate addition reactions with nucleophiles (eqn 7.18). A powerful driving force for this reaction is provided by enolisation of the initial product to give the fully aromatic quinol.

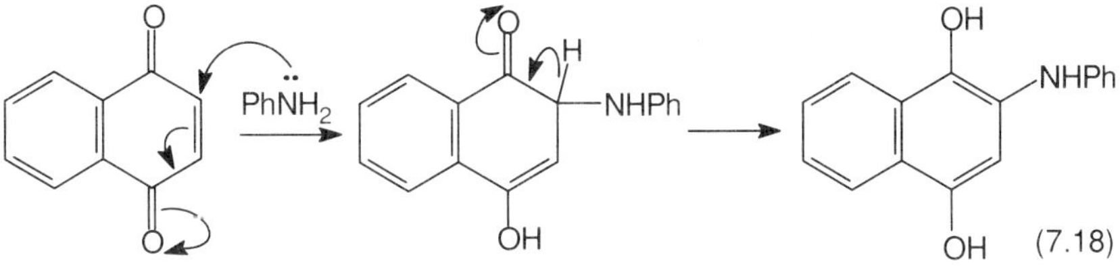

quinone   quinol

(7.18)

Quinones are important because of their widespread occurrence in nature as products of plant and animal metabolism. Thus, vitamin K_1 is found in green plants and plays a vital role in maintaining the coagulent properties of blood.

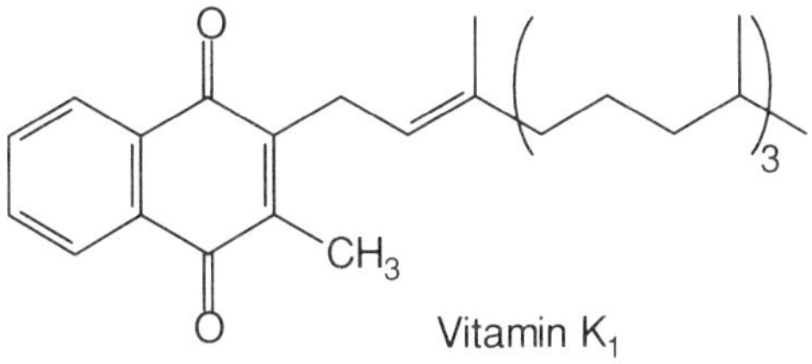

Vitamin K_1

7.4 βγ- and γδ-unsaturated carbonyl compounds

Non-conjugated unsaturated acids undergo electrophile-induced cyclisation reactions such as iodolactonisation (e.g. eqns 7.19 and 7.20). Such reactions involving βγ- and γδ-unsaturated acids invariably yield the five-membered lactone ring.

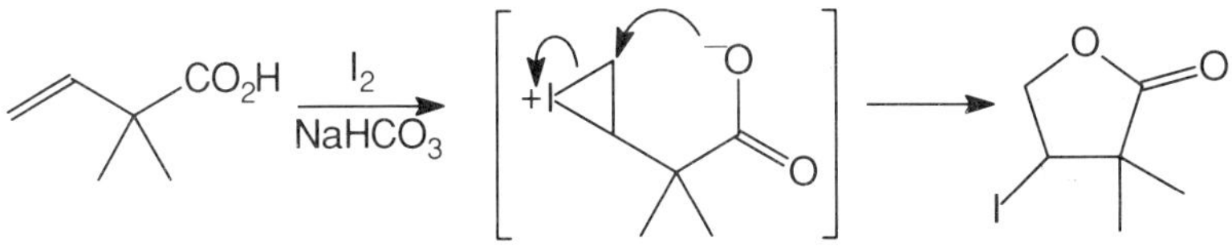

(7.19)

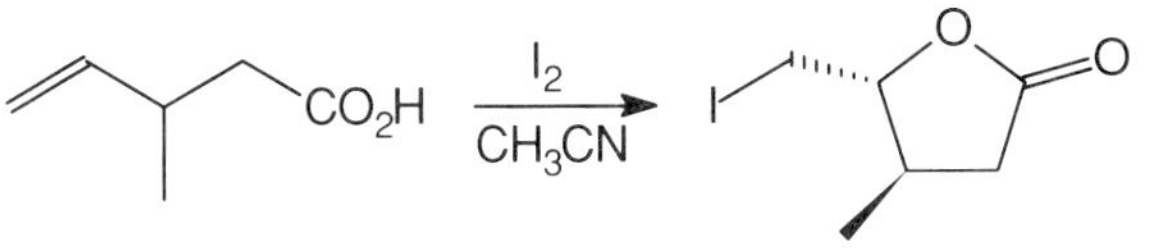

(7.20)

Problems

1. Suggest structures for compounds **A** and **B** in the following scheme.

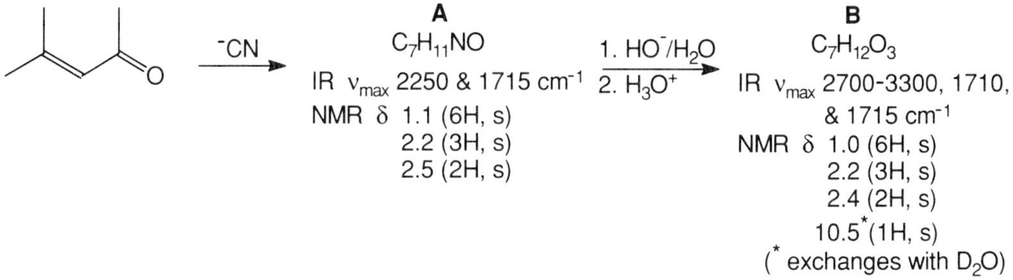

2. Suggest possible mechanisms for each of the following reactions.

(a)

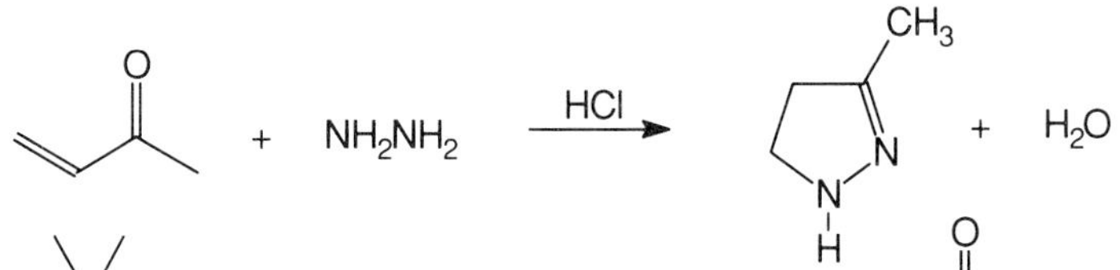

(b)

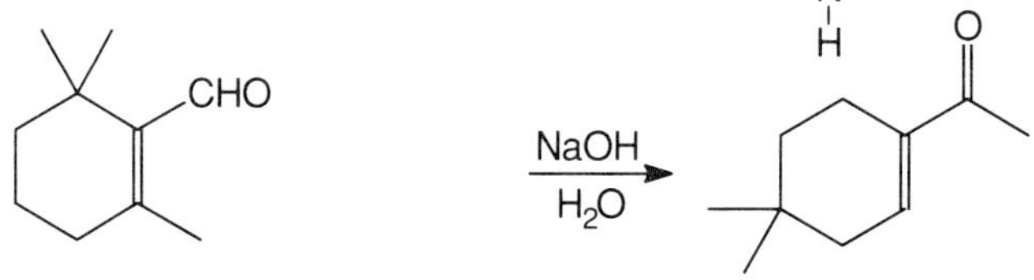

(c)

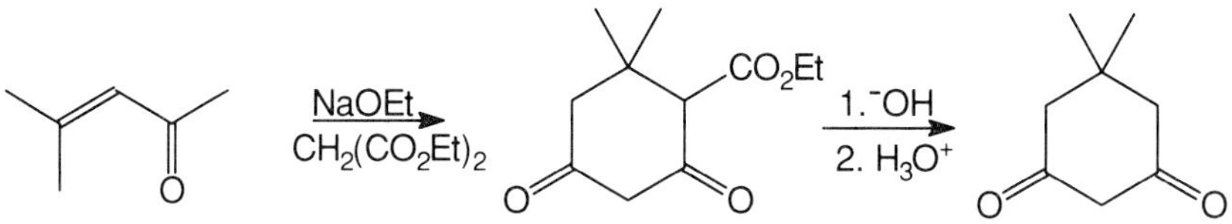

(d)

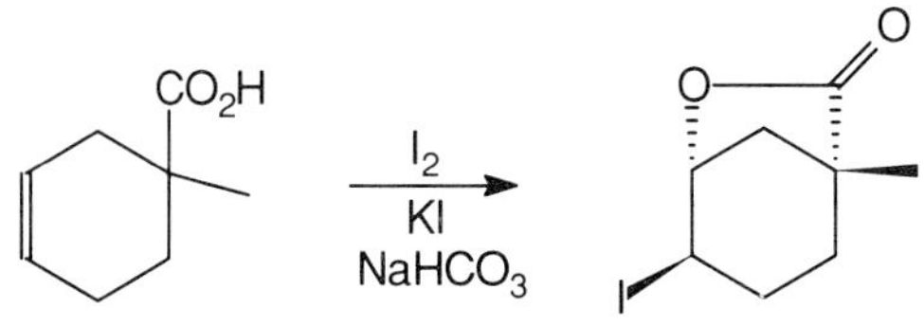

8 Enamines, enol ethers, and enolates

8.1 Enamines

Enamines are nucleophilic reagents and as such they undergo alkylation, acylation and conjugate addition reactions (e.g. eqns 8.1–8.3). Indeed in some situations they have advantages over other reagents for bringing about such reactions. For example, since a strong base is not required, wasteful self-condensation reactions are precluded. Furthermore, since enamines themselves are prepared from aldehydes and ketones (see section 2.5), and since their salts are hydrolysed by aqueous acid to regenerate the carbonyl group, they afford an indirect route for preparing substituted carbonyl compounds.

(8.1) Alkylation

(8.2) Acylation

(8.3) Michael addition

8.2 Enol ethers and enolates

For most aldehydes and ketones the enol tautomer is less stable than the keto form with the result that very little of the enol is present. However, for some carbonyl compounds (e.g. β-diketones) the enol tautomer predominates, at least in some solvents, due to extra stabilisation afforded by intramolecular hydrogen bonding (see section 6.2). For any carbonyl compound, interconversion of the keto and enol forms is catalysed by acid or base (eqns 8.4 and 8.5), and the enolate anions formed by deprotonating carbonyl compounds are stabilised by resonance.

$$(8.4)$$

$$(8.5)$$

Stoichiometric amounts of enolate anions can be generated by treating carbonyl compounds with an equivalent amount of a strong base. They are very useful in organic synthesis since they react with electrophiles such as alkyl halides leading to carbon–carbon bond formation. However, complications such as dialkylation and poor regioselectivity can arise when direct alkylation of enolate anions is attempted (e.g. eqns 8.6 and 8.7).

A *regioselective* reaction is one which produces unequal amounts of two (or more) structurally isomeric products.

$$(8.6)$$

$$(8.7)$$

It is for this reason that β-ketoesters and enamines have proved so valuable for preparing substituted carbonyl compounds. Fortunately, however, there are now a number of indirect ways in which enolates and their equivalents can be generated regioselectively so that, in many cases, these difficulties can be avoided (e.g. eqns 8.8 and 8.9)

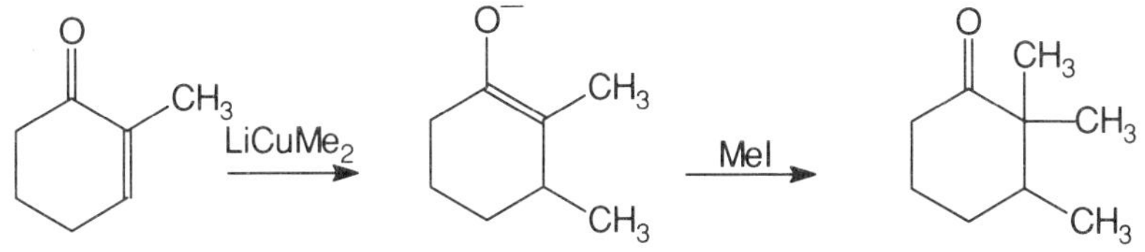

(8.8)

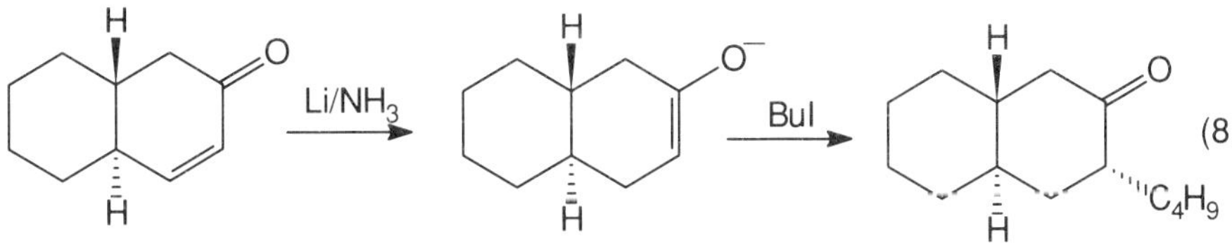

(8.9)

Furthermore, the factors influencing the geometry and reactivity of enolate derivatives have been extensively studied and considerable control can now be exercised over direct enolate alkylation reactions. For example, it is known that the preferred enolate geometry depends upon the structure of the substrate and the solvent used for the deprotonation reaction (see scheme 8.1). One way of obtaining this information is to trap the enolate as its silyl enol ether by reaction with trimethylsilyl chloride (see section 2.5).

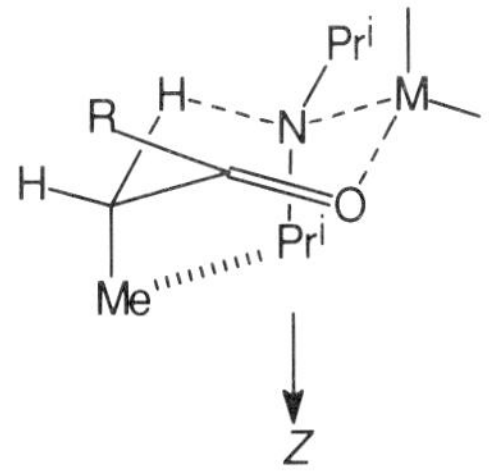

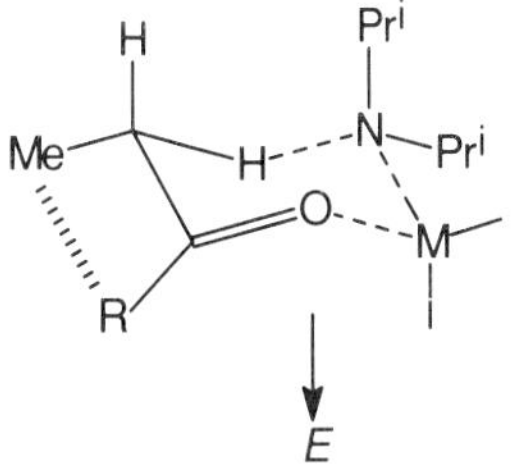

R =	Et	Pr^i	Bu^t
Z	30	60	>98
E	70	40	<2

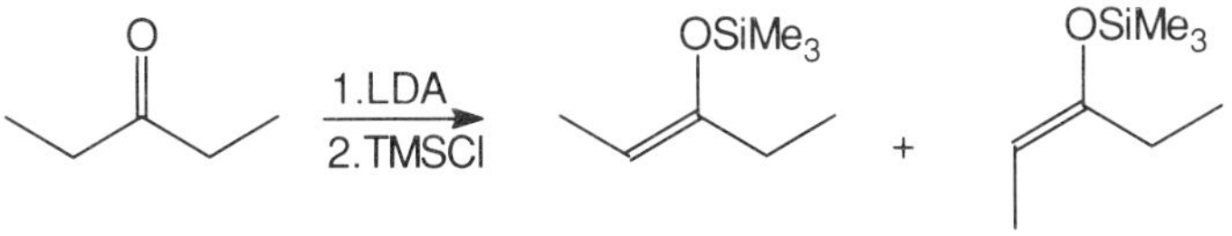

	Z	E
THF	30	70
THF/HMPA	82	18

Scheme 8.1   HMPA = (Me$_2$N)$_3$PO

When R is small the E enolate predominates (when M = Li), but when R is large the Z enolate is

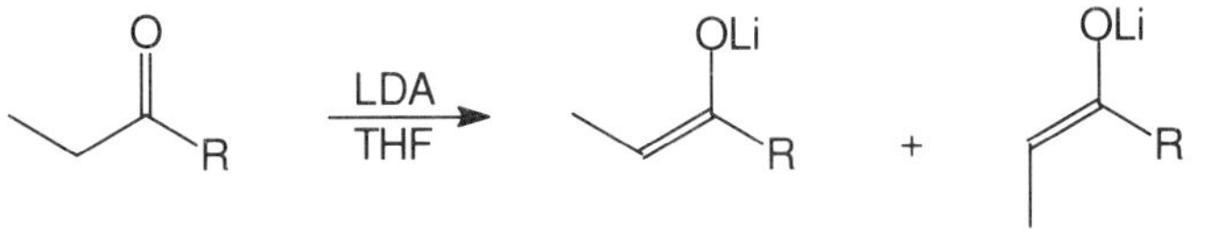

By controlling the enolate geometry and by using a chiral auxiliary group, highly diastereoselective alkylation reactions can be carried out as in scheme 8.2 (d.e.= excess of major diastereoisomer). Furthermore, by using other electrophiles the asymmetric synthesis of a wide variety of α-substituted carbonyl compounds can be achieved (scheme 8.3).

A *stereoselective* reaction is one which produces unequal amounts of two (or more) stereoisomeric products (in the case of a *diastereoselective* reaction the products formed are diastereoisomers).

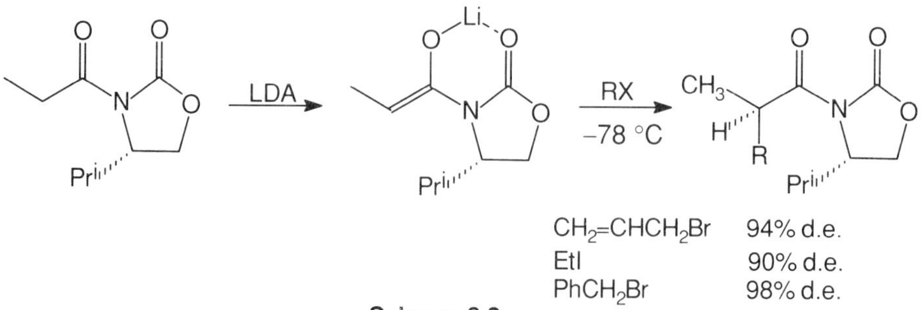

Scheme 8.2

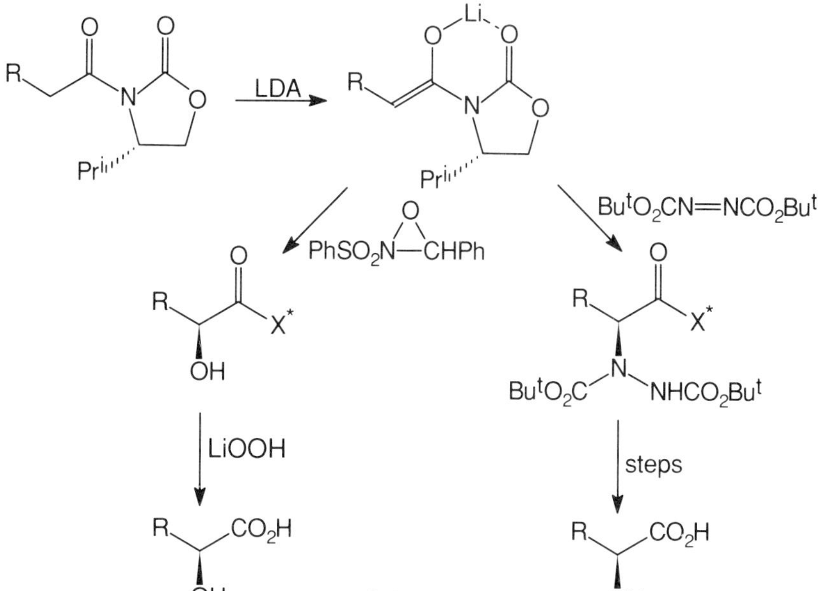

CH$_2$=CHCH$_2$Br   94% d.e.
EtI   90% d.e.
PhCH$_2$Br   98% d.e.

Scheme 8.3

One enolate reaction which has been extensively studied is the aldol reaction. In this case the preferred enolate geometry determines the overall diastereoselectivity (i.e. *syn/anti* ratio) of the reaction (see section 2.3).

A further limitation with many base-catalysed reactions is the ease with which secondary and tertiary alkyl halides undergo elimination reactions (e.g. eqn 8.12).

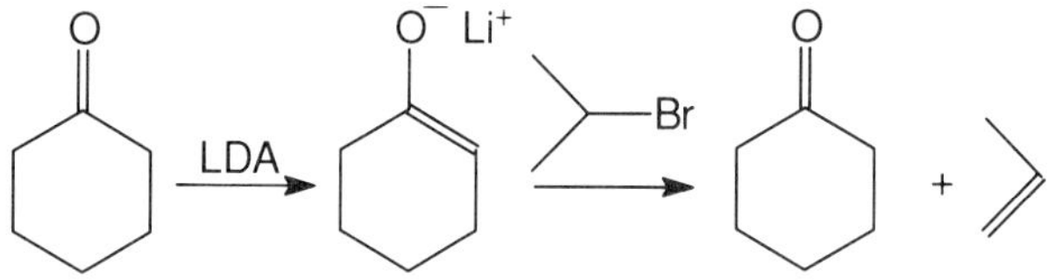

The use of a silyl enol ethers under Lewis acid catalysed conditions affords a useful method for carrying out alkylation reactions in such cases (e.g. eqns 8.13 and 8.14).

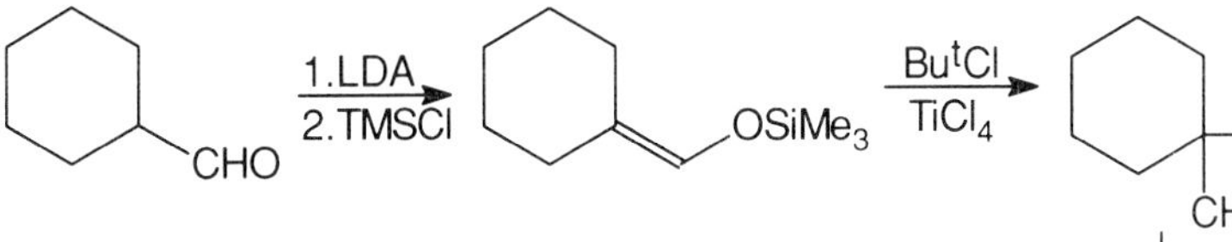

(8.13)

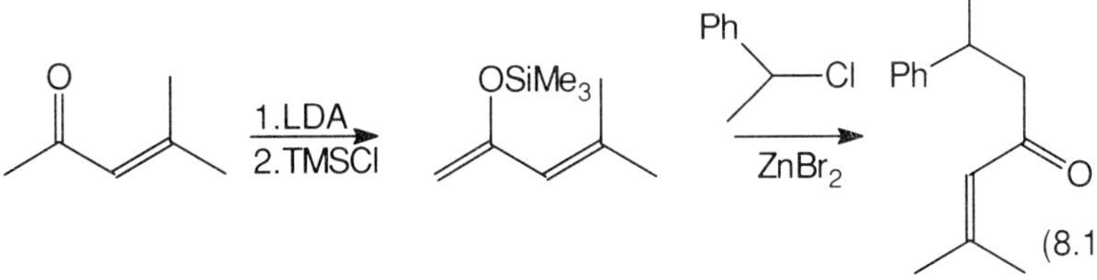

(8.14)

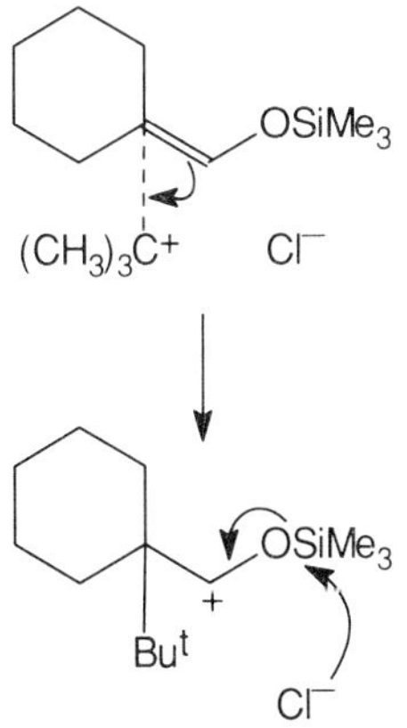

Silyl enol ethers, which can be regioselectively generated and purified, provide an alternative way of bringing about regioselective reactions on unsymmetrical ketones (eqns 8.15 and 8.16).

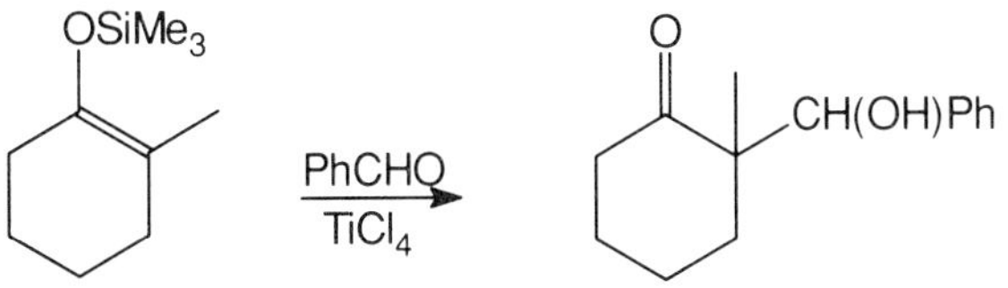

(8.15)

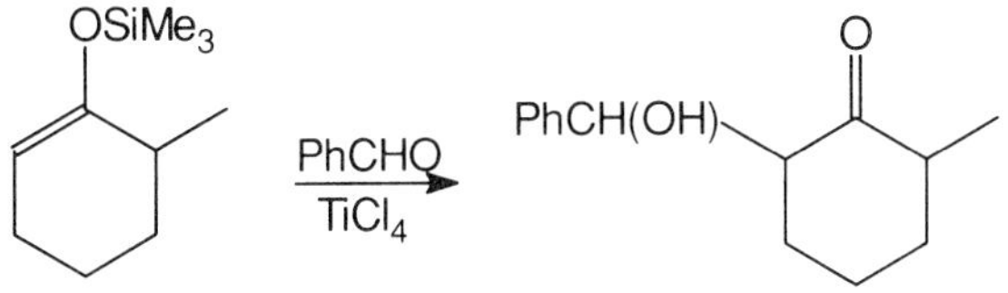

(8.16)

9 Allyl compounds

9.1 Allyl alcohols

Allyl (and benzyl) alcohols are renowned for their ease of oxidation. Manganese dioxide is particularly effective for bringing about oxidation of primary allylic alcohols to the $\alpha\beta$-unsaturated aldehydes (e.g. eqn 9.1).

vitamin A   retinal   (9.1)

A second highly-selective oxidation of allylic alcohols involves epoxidation of the carbon–carbon double bond using *tert*-butyl hydroperoxide and a titanium or vanadium catalyst (e.g. eqn 9.2). This is the basis of an extremely valuable method of asymmetric synthesis, known as *Sharpless epoxidation* which involves treating the allylic alcohol with *tert*-butyl hydroperoxide in the presence of titanium tetraisopropoxide and a chiral ester of tartaric acid (e.g. eqns 9.3 and 9.4). The products are epoxy alcohols which can be converted into a wide variety of other products by reduction or nucleophilic addition (see section 2.2). Since the reaction is highly stereoselective it provides a useful approach to preparing compounds with high enantiomeric purity.

This is an example of an *enantioselective* reaction since the two possible products are enantiomers.

DET = diethyl tartrate

9.2 Allyl ethers

Allyl vinyl ethers (and allyl phenyl ethers) undergo a [3,3] sigmatropic rearrangement on heating, analogous to the Cope rearrangement of 1,5-dienes (see section 3.5). The reaction is known as the *Claisen rearrangement* (scheme 9.1), and like the Cope rearrangement, it proceeds via a chair-like transition state in which substituent groups adopt, whenever possible, an equatorial orientation. When both ends of the allyl vinyl ether are substituted, two new stereogenic centres are created and diastereoisomeric products are selectively obtained.

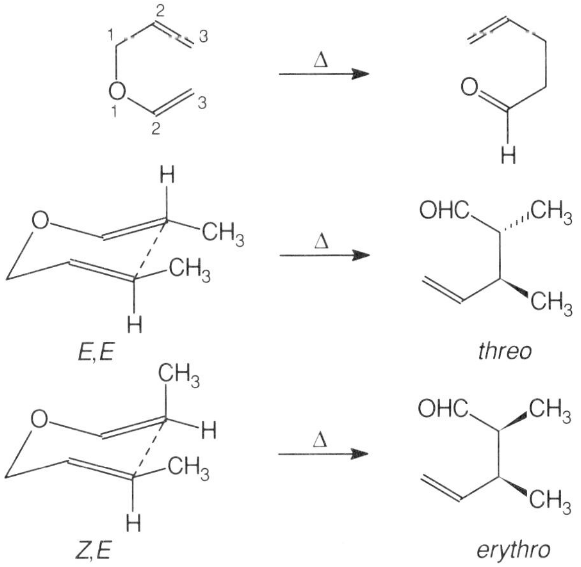

Scheme 9.1

When both of the *ortho* positions of an allyl phenyl ether are blocked a second (Cope) rearrangement occurs leading eventually to the *para* substituted allyl phenol (eqns 9.5 and 9.6).

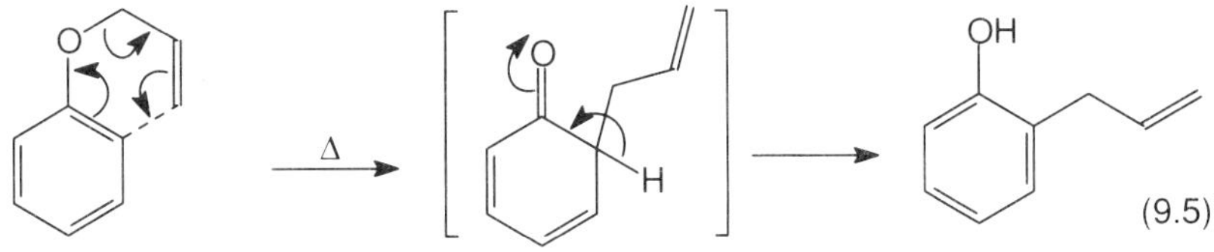

$$(9.5)$$

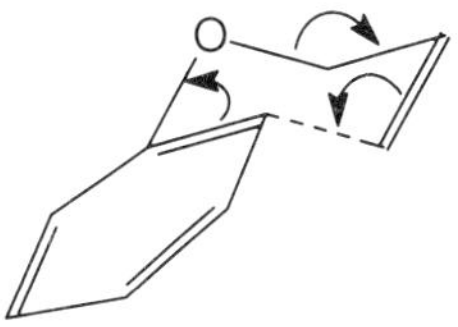

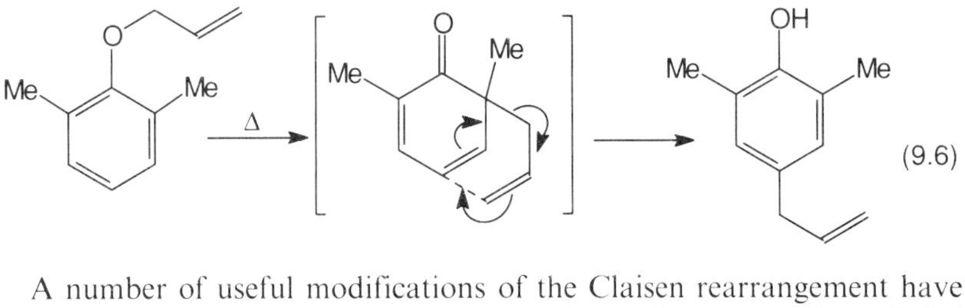

$$(9.6)$$

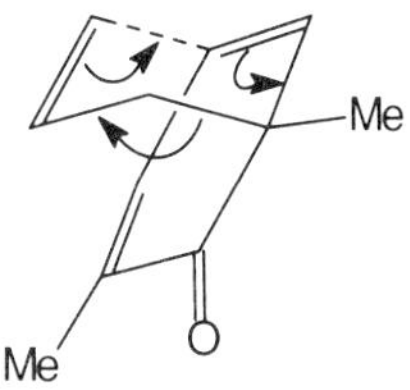

A number of useful modifications of the Claisen rearrangement have been introduced with the result that it can be applied to a variety of allyl alcohol derivatives and provides a valuable method for carbon–carbon bond formation (eqns 9.7–9.9).

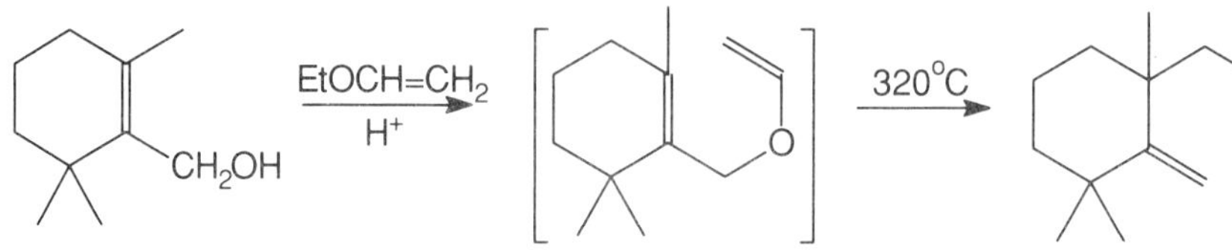

$$\text{(9.7)}$$

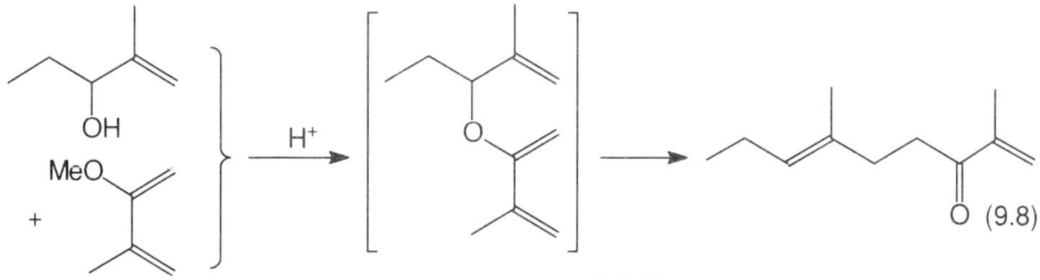

$$\text{(9.8)}$$

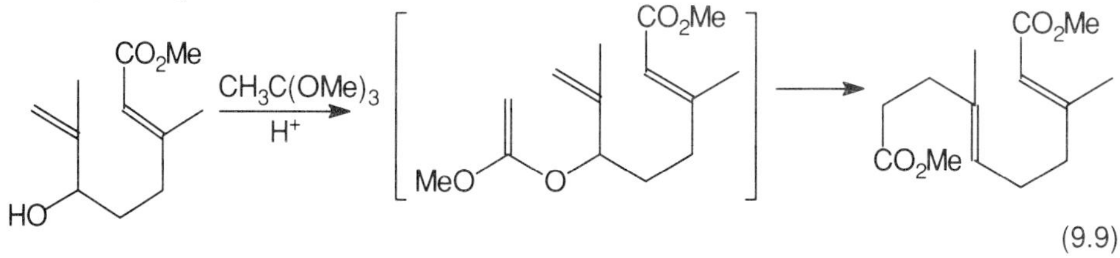

$$\text{(9.9)}$$

9.3 Allyl halides

The ready ionisation of allyl (and benzyl) halides has important consequences. For example, hydrolysis of either 1-chloro-2-butene or 3-chloro-1-butene proceeds via a unimolecular mechanism and gives the same mixture of allyl alcohols (scheme 9.2). The reason for this is the formation of the same allylic cation.

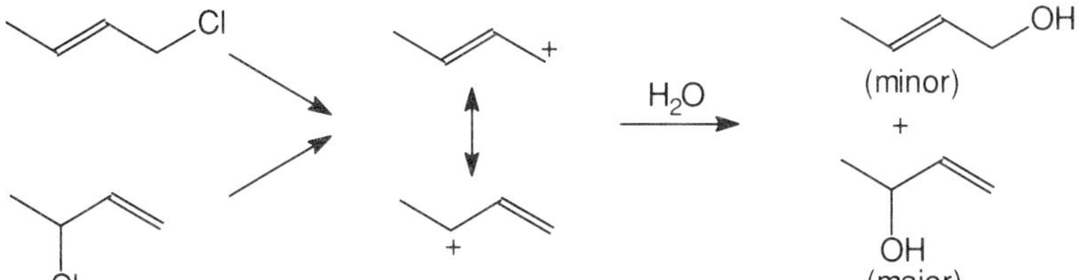

(minor)

(major)

Scheme 9.2

Allyl halides can also undergo S_N2 reactions, and react faster than the corresponding primary alkyl halides. Such bimolecular displacement processes can however be complicated by the occurrence of the S_N2' reaction in which substitution is accompanied by movement of the carbon–carbon double bond (e.g. eqn 9.10).

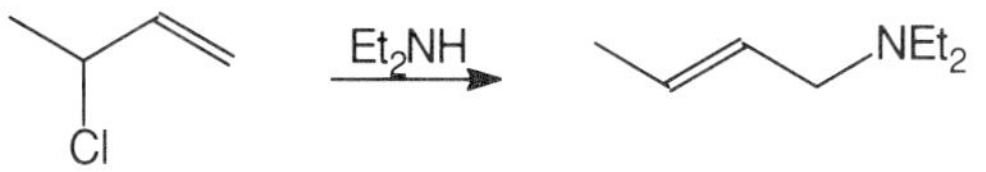

$$\xrightarrow{\text{Et}_2\text{NH}} \qquad \text{(9.10)}$$

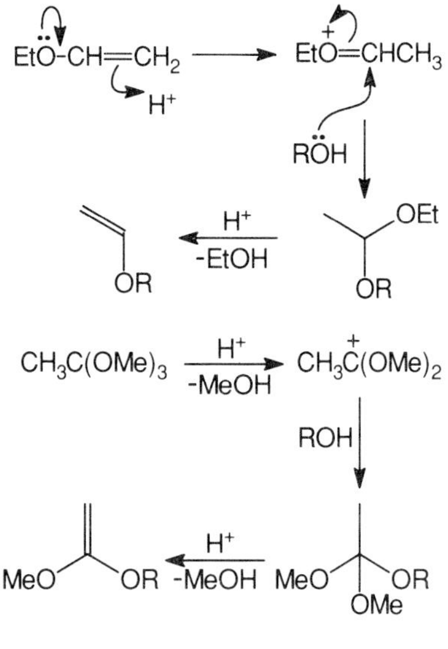

Allyl and benzyl halides are readily ionised because the carbocations produced are stabilised.

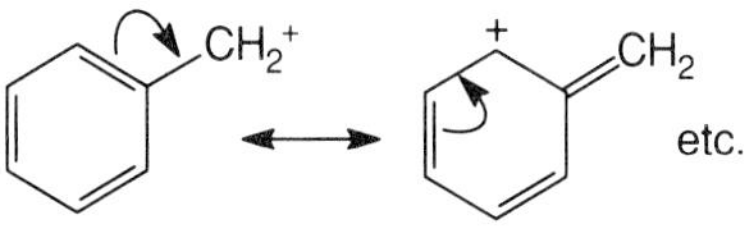

a benzyl cation

10 Selective protection of bifunctional compounds

We have seen that certain bifunctional compounds are extremely useful in organic synthesis. Nevertheless, the presence of two or more reactive functional groups in a molecule can sometimes be more of a hindrance than a help. When a molecule contains two reactive functional groups, the problem which often arises is how to carry out a selective reaction on one of them without affecting the other. Three different situations can arise.

1. If the two groups differ in their reactivity then it should be possible to perform a reaction on the more reactive group (eqn 10.1).

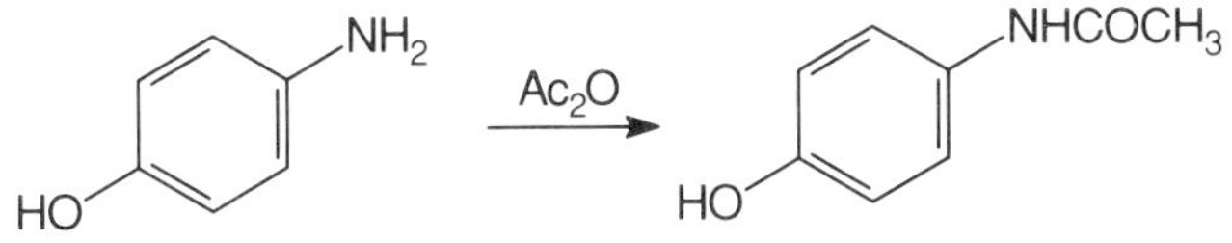

(10.1)

2. If the two groups are identical but the product of reaction is less reactive than the starting material then once again it should be possible to bring about a selective reaction (eqns 10.2 and 10.3).

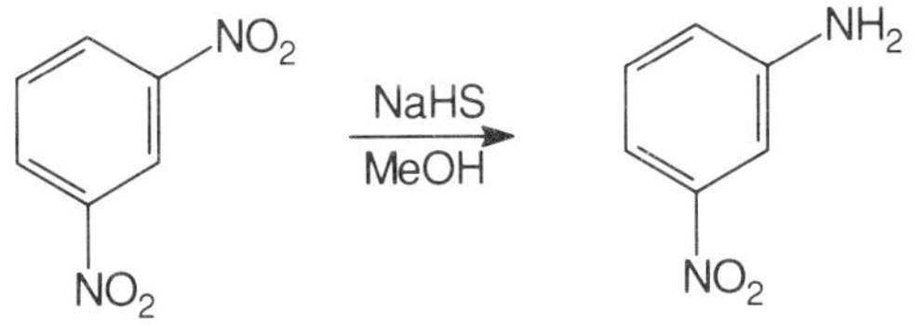

(10.2)

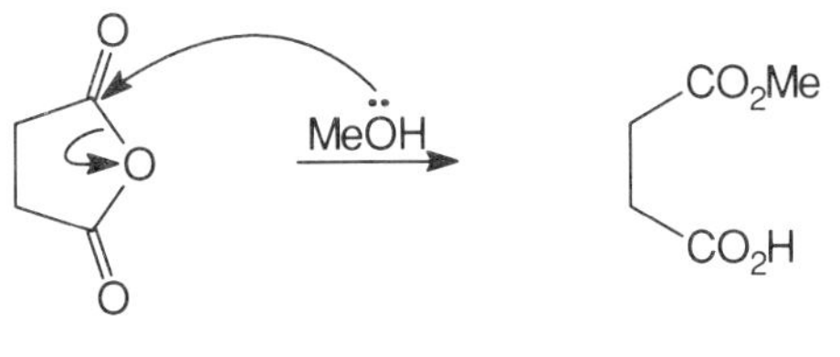

(10.3)

3. However, if we wish to carry out a reaction at the less reactive of two functional groups then we will usually need to protect the more reactive group (scheme 10.1).

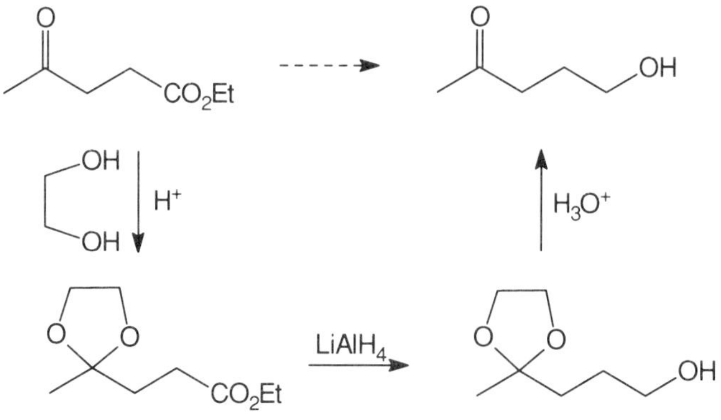

Scheme 10.1

10.1 Use of protecting groups

As illustrated above, acetals can be used to protect aldehyde or ketone groups while transformations are carried out elsewhere in a molecule. Acetals can also be used to protect diols and simple alcohols. Indeed they are often preferable to simple ether protecting groups because they can be removed under very mild conditions. Thus, the so-called tetrahydro-pyranyl (THP) and methoxymethyl (MOM) ethers (eqns 10.4 and 10.5), are in reality acetals and can be readily removed by treatment with aqueous acid.

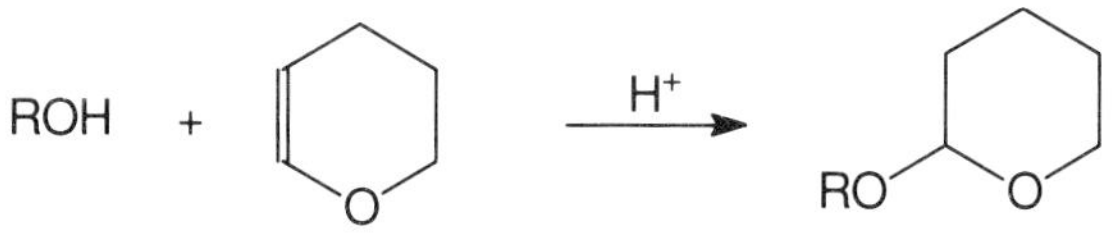

(10.4)

(THP ether)

(10.5)

(MOM ether)

A short synthesis illustrating the use of the THP protecting group is shown in scheme 10.2.

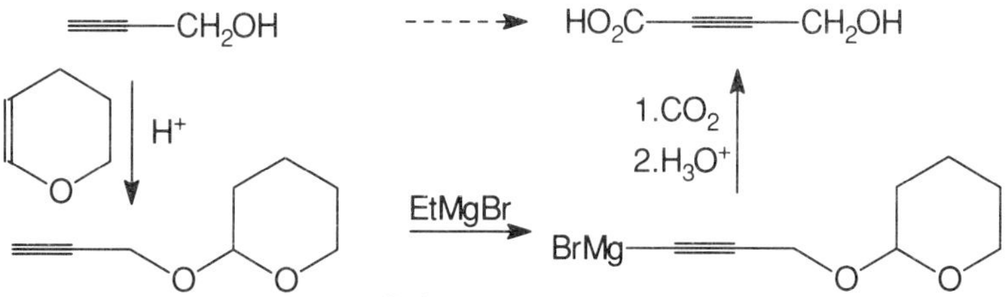

Scheme 10.2

Alcohols can also be protected as trimethylsilyl ethers (eqn 10.6). The trimethylsilyl protecting group can be removed by treatment with tetrabutylammonium fluoride or aqueous HF.

$$ROH \; + \; Me_3SiCl \quad \xrightarrow{R_3N} \quad ROSiMe_3 \qquad (10.6)$$

$$\text{(TMS ether)}$$

10.2 Selective protection/deprotection of carbohydrates

One area where alcohol (and carbonyl) protecting groups have been widely exploited is in the chemistry of carbohydrates since in any synthesis involving carbohydrates one is immediately faced with the problem of distinguishing between several very similar functional groups. Scheme 10.3 illustrates the use of acetal and ether groups to prepare two selectively methylated glucose derivatives.

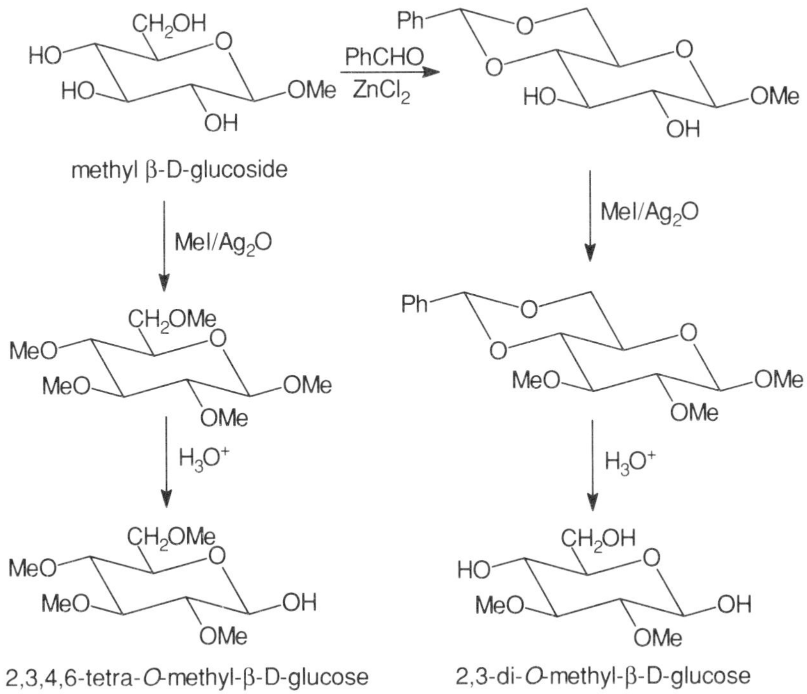

methyl β-D-glucoside

2,3,4,6-tetra-*O*-methyl-β-D-glucose   2,3-di-*O*-methyl-β-D-glucose

Scheme 10.3

10.3 Selective protection/deprotection of amino acids

Peptide synthesis depends upon the use of carboxyl and amino protecting groups, since it involves joining together different monomer units containing carboxyl and amino groups. Furthermore, the protecting groups must be introduced and removed under conditions which do not cleave peptide bonds and do not lead to racemisation.

Carboxyl protection

The use of methyl or ethyl esters as protecting groups is clearly not satisfactory since the strong alkali or aqueous acid needed to remove the alkyl group would lead to cleavage of the amide bonds and epimerisation of the stereogenic centres. Esters which can be cleaved without base treatment are therefore ideal protecting groups for this purpose. Benzyl esters which can be cleaved by hydrogenolysis, and *tert*-butyl esters which can be removed by treatment with anhydrous acid (dry HCl or tri-fluoroacetic acid) are usually used (eqns 10.7 and 10.8).

$$\text{(10.7)}$$

$$\text{(10.8)}$$

Amino protection

The use of a conventional acyl derivative (amide) as an amine protecting group suffers from the disadvantage that it would again require the use of strong alkali or aqueous acid to remove it. The benzyloxycarbonyl and *tert*-butoxycarbonyl (*t*-Boc) groups have therefore been developed for this purpose, since they can be readily removed under conditions which do not harm the peptide product (eqns 10.9 and 10.10).

$$\text{(10.9)}$$

$$\text{(10.10)}$$

Peptide formation

Peptide synthesis involves not only protecting the functional groups not taking part in the reaction but also activating the carboxyl group to bring about amide bond formation (eqn 10.11). However, activation of the carboxyl group enhances the possibility of epimerisation by increasing the acidity of the α-hydrogen atom, and may also cause other side reactions to occur.

$$\text{(10.11)}$$

W = amino protecting group
X = carboxyl protecting group
Y = activating group

Three methods which have been successfully used to make peptide bonds are illustrated below. The first involves converting the carboxyl group into an acyl azide (eqn 10.12). These derivatives are more useful than acyl halides since, for reasons which are not clear, they do not undergo racemisation.

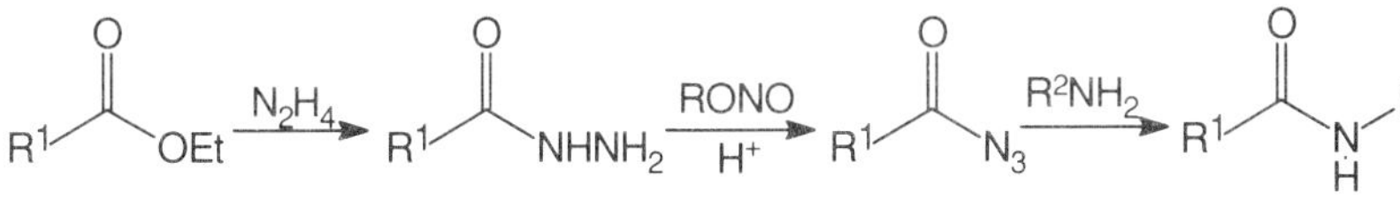

$$(10.12)$$

A second approach involves the use of an activated ester derived from a phenol, such as pentafluorophenol or 2,4,5-trichlorophenol, the anion of which provides a good leaving group. This approach is illustrated by a synthesis of the artificial sweetener aspartame (scheme 10.4).

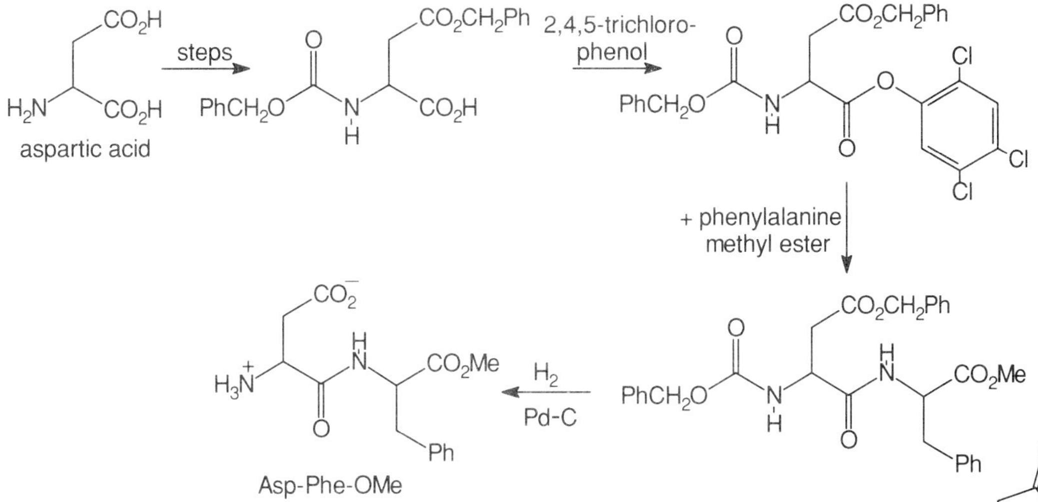

Scheme 10.4

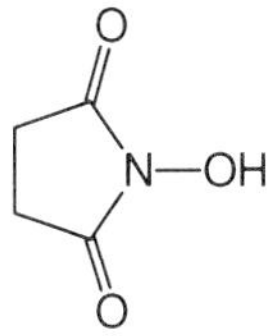

The third method uses dicyclohexylcarbodiimide (DCCI), which has the effect of converting the carboxylic acid into an activated intermediate which is not isolated (eqn 10.14). The success of this method depends upon the fact that a thermodynamically stable urea molecule is liberated when the intermediate reacts with an amine. In practice DCCI is usually used in conjunction with *N*-hydroxysuccinimide or *N*-hydroxybenzotriazole, which facilitates the coupling process while at the same time minimising the risk of racemisation.

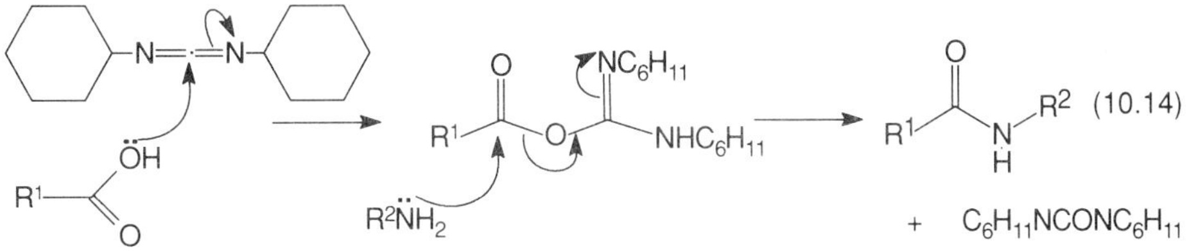

N-hydroxysuccinimide

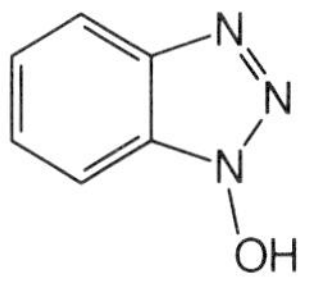

1-hydroxybenzotriazole

$$(10.14)$$

$$+ \quad C_6H_{11}NCONC_6H_{11}$$

A short dipeptide synthesis using DCCI is shown in scheme 10.5.

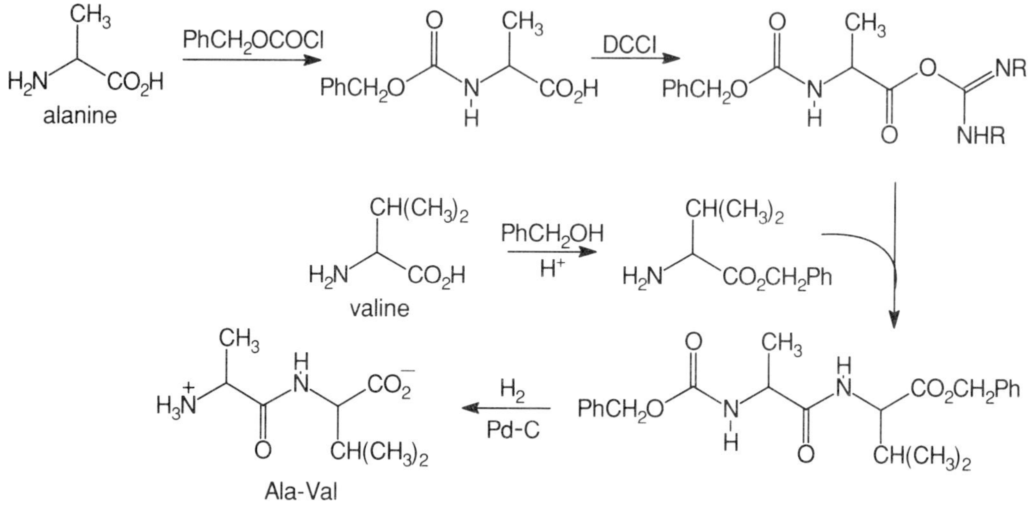

Scheme 10.5

Problems

1. Outline methods by which each of the following compounds may be synthesised from the suggested starting material.

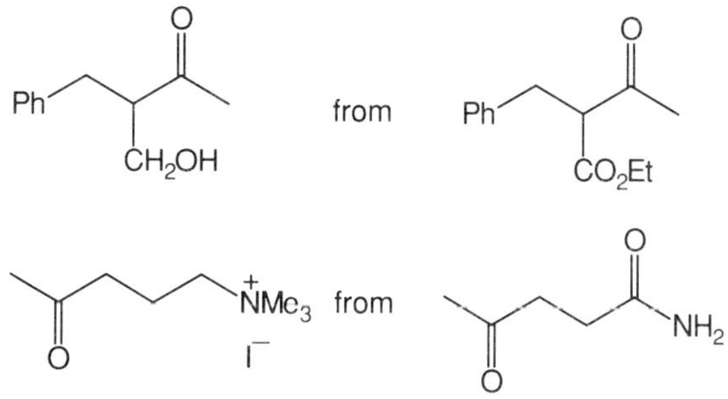

2. Suggest structures for the compounds **A–D** in the following scheme.

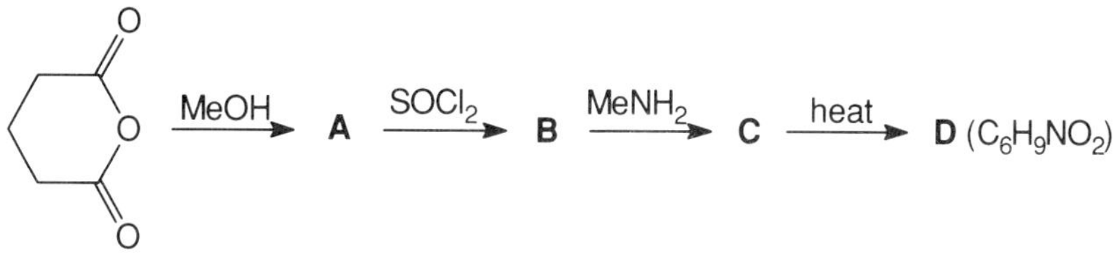

11 Cyclisation versus polymerisation

As indicated in Chapter 1, many bifunctional compounds are ideally suited to undergo either an intramolecular reaction leading to cyclic products or an intermolecular reaction leading to polymeric products. The competition between cyclisation and polymerisation can be influenced in several ways, the most obvious being to vary the concentration of the solution. Thus one method which can be used to promote cyclisation at the expense of polymerisation is to carry out the reaction in dilute solution so that the probability of an intermolecular reaction taking place is reduced.

In general terms the ease of ring closure is determined by two factors.

1. An entropy factor. This can be regarded as a measure of the probability of the two ends of the chain coming into contact with one another. To a first approximation this decreases as the chain length increases.

2. An enthalpy factor. This can be correlated with the stability of the ring being produced and is most favourable for six-membered rings. An estimate of the relative stability of alicyclic rings can be obtained by comparing their heats of combustion per CH_2 group (Table 11.1).

Table 11.1 Heats of combustion per CH_2 for cycloalkanes

Ring size	Heat of combustion per CH_2 kJ mol^{-1}	Ring size	Heat of combustion per CH_2 kJ mol^{-1}
3	699.72	10	666.12
4	688.80	11	665.28
5	666.54	12	661.92
6	661.08	13	662.76
7	664.86	14	661.08
8	666.12	15	661.50
9	666.96	17	660.24

Taking both factors into account leads to the conclusion that five- and six-membered rings will be easily obtained by cyclisation, whereas three- and four-membered rings will be more difficult to obtain by this method. Indeed, special methods including cycloaddition reactions are often employed to prepare small ring compounds.

The relative ease of cyclisation of a series of ω-bromoalkylamines are illustrated in scheme 11.1.

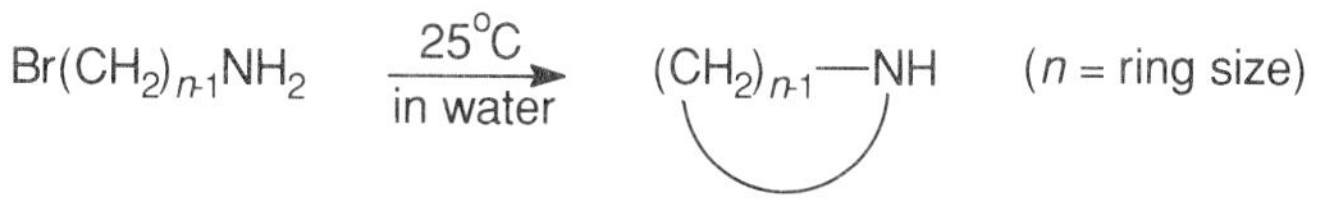

$$Br(CH_2)_{n\text{-}1}NH_2 \xrightarrow[\text{in water}]{25°C} (CH_2)_{n\text{-}1}\!\!-\!\!NH \qquad (n = \text{ring size})$$

$n = 3$	rel.rate $= 0.12$
4	0.002
5	100.00
6	1.67
7	0.03
10	10^{-8}

Scheme 11.1

A further consideration which affects the course of many cyclisation reactions is the nature of the electrophilic centre being attacked. Ring closure is favoured when the terminal atoms can achieve the correct geometry for reaction. This aspect is summarised by a set of empirical rules known as *Baldwin's rules*. These classify reactions according to the size of the ring being produced, whether the group Y lies outside the ring or becomes part of the ring itself (*exo* or *endo*), and the geometry (e.g. tetrahedral or trigonal) of the carbon atom being attacked (scheme 11.2). The rules apply to the formation of three- to seven-membered rings and predict, for example, that *Exo-Tet* and *Exo-Trig* processes are favoured for all ring sizes, but *Endo-Trig* processes are only favoured for six- and seven-membered rings. Furthermore 5- and 6-*Endo-Tet* processes and 3- to 5-*Endo-Trig* processes are particularly disfavoured.

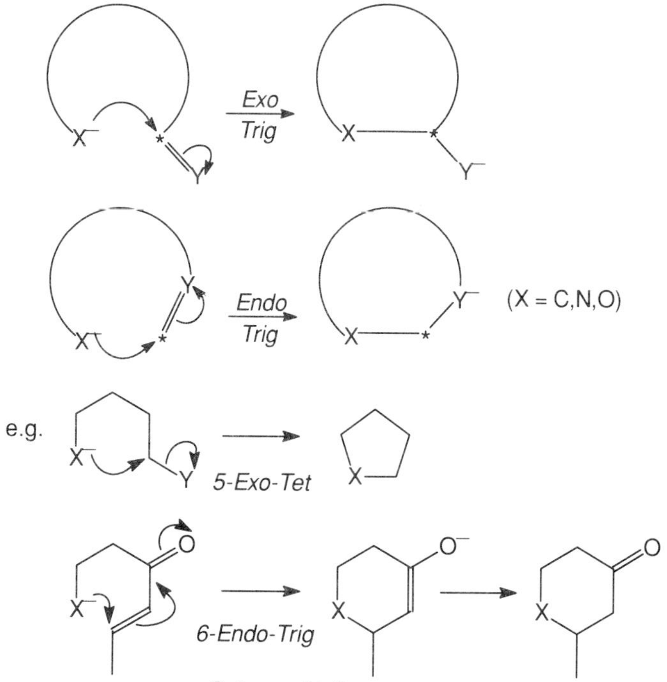

Scheme 11.2

11.1 Preparation of non-aromatic carbocyclic compounds

In the preceding chapters we have encountered several reactions which can be used to synthesise carbocyclic compounds. The most important of these are listed below and typical examples are illustrated (eqns 11.1–11.6).

Acyloin condensation

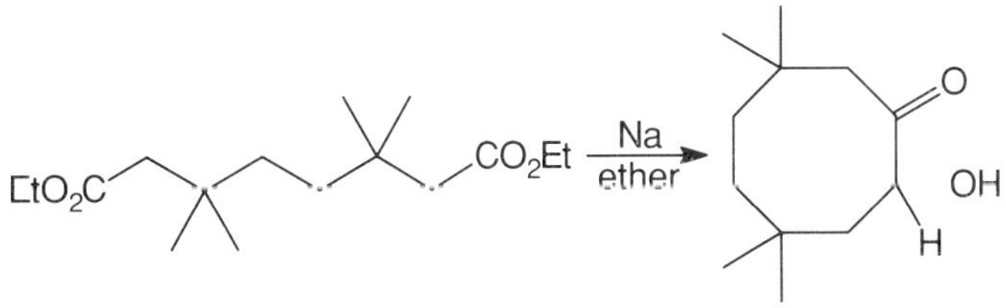

$$(11.1)$$

Aldol condensation

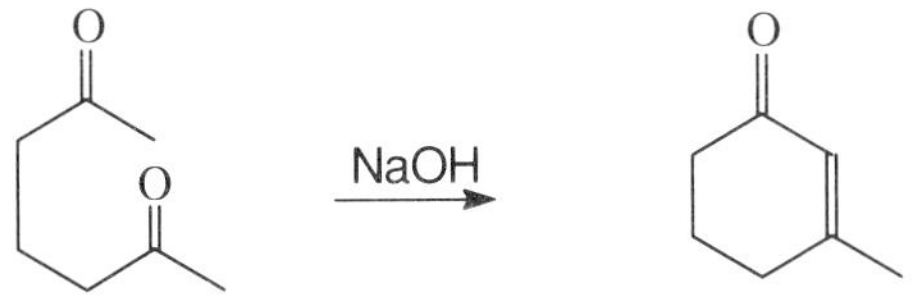

$$(11.2)$$

Claisen (Dieckman) condensation

$$(11.3)$$

Alkylation of β-dicarbonyl compounds

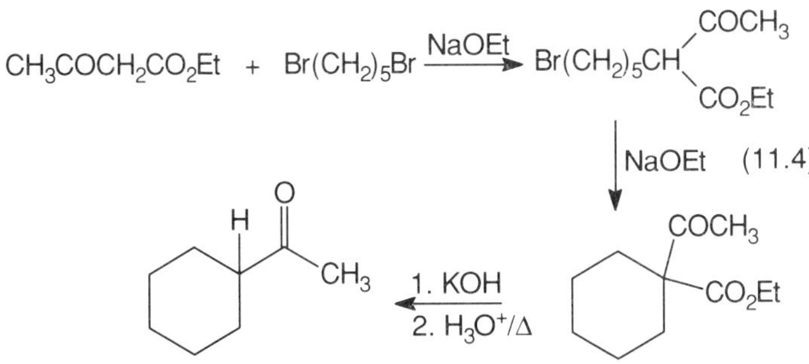

$$(11.4)$$

Robinson annelation

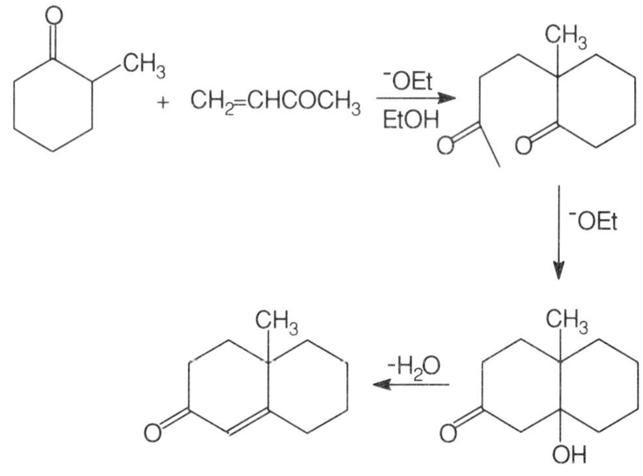

$$\text{(11.5)}$$

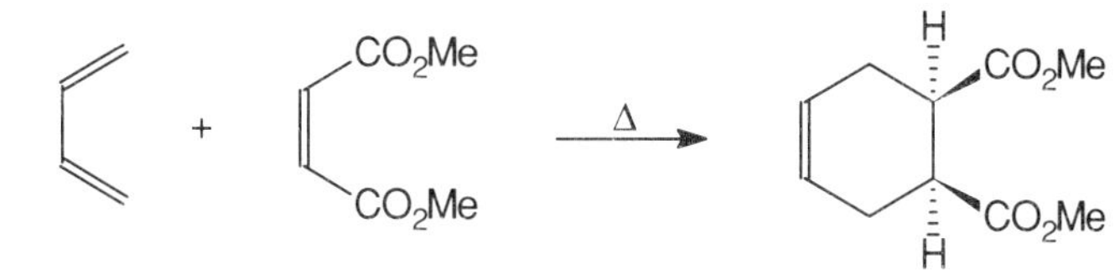

$$\text{(11.6)}$$

Diels–Alder

11.2 Preparation of non-aromatic heterocyclic compounds

Bifunctional compounds are also widely used for the preparation of heterocyclic compounds. The synthesis of aromatic heterocyclic compounds is the subject of a separate primer in this series. Some of the methods used for the synthesis of small (three- to six-membered) non-aromatic heterocyclic compounds are outlined here. This is not however intended to be a comprehensive review of the methods available, but rather to give examples of some of the ingeneous approaches which can be adopted involving bifunctional compounds.

Three-membered rings

Oxiranes (epoxides) can be prepared by intramolecular cyclisation of halohydrins under basic conditions (eqns 11.7 and 11.8).

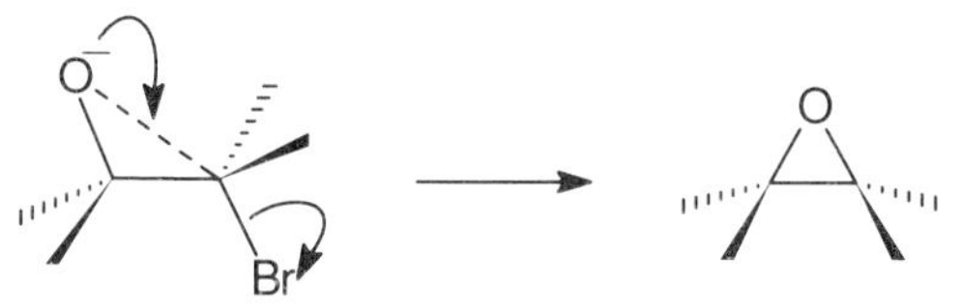

$$\text{(11.7)}$$

$$(11.8)$$

Oxiranes can also be prepared by addition of hydroperoxide to αβ-unsaturated ketones (eqn 11.9).

$$(11.9)$$

However the most common method for the preparation of oxiranes is by epoxidation of alkenes using, for example, a peracid (eqn 11.10). A similar reaction can be used to convert alkenes into aziridines (eqn 11.11).

$$+ RCO_3H \quad \text{a peracid} \qquad (11.10)$$

See page 8 for the mechanism of epoxide formation.

$$:NCO_2Me \quad \text{a nitrene} \qquad \qquad NCO_2Me \quad (11.11)$$

Aziridines and thiiranes can also be prepared by methods involving cyclisation (eqns 11.12 and 11.13).

$$H_2NCH_2CH_2Cl \xrightarrow{^-OH} \quad \triangleright NH \qquad (11.12)$$

$$\xrightarrow[-5^\circ C]{KSCN(aq)} \qquad \xrightarrow{-OCN^-} \qquad (11.13)$$

Four-membered rings

Oxetanes, azetidines and thietanes can all be prepared by cyclisation methods but the yields are usually only modest (eqn 11.14).

$$\xrightarrow[Et_2O]{NaH} \qquad (55\%) \quad (11.14)$$

Better yields, at least in the case of oxetanes, are usually obtained by using cycloaddition reactions (e.g. eqn 11.15).

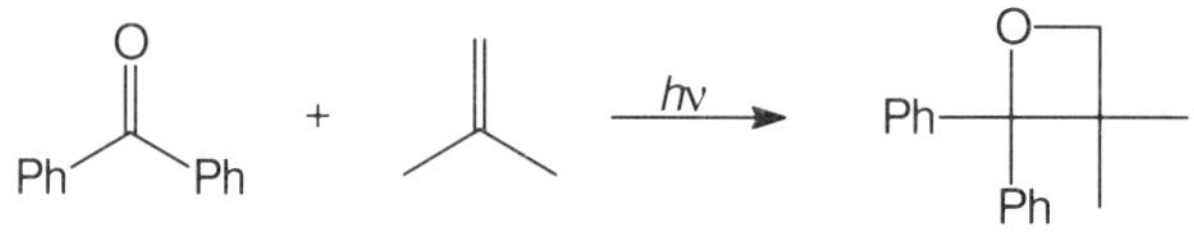

$$\text{(11.15)} \quad \text{(93\%)}$$

Five-membered rings

The ready cyclisation of 1,4-diols, 1,4-diketones and γ-hydroxycarbonyl compounds has been covered elsewhere (see sections 4.3, 5.1, 5.2, and 6.3). Other five-membered heterocycles can be prepared using simple addition/condensation reactions of β-dicarbonyl compounds and αβ-unsaturated carbonyl compounds (e.g. eqns 11.16 and 11.17).

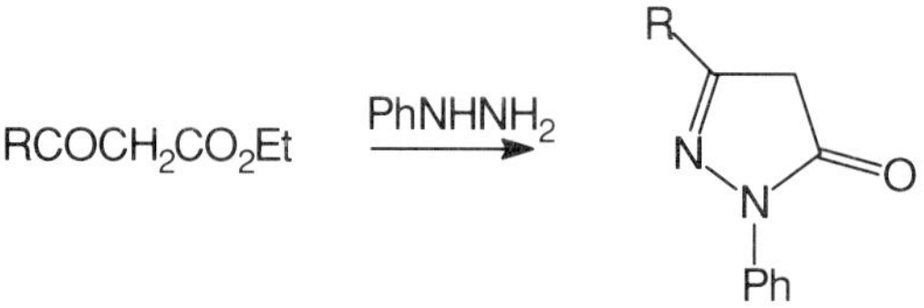

$$\text{RCOCH}_2\text{CO}_2\text{Et} \xrightarrow{\text{PhNHNH}_2} \qquad \text{(11.16)}$$

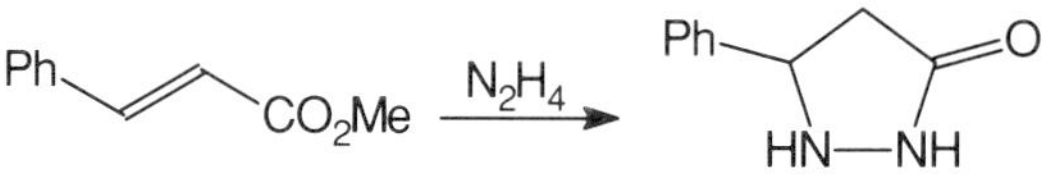

$$\text{Ph} \diagup \text{CO}_2\text{Me} \xrightarrow{\text{N}_2\text{H}_4} \qquad \text{(11.17)}$$

It should however be remembered that many heterocyclic compounds can also be prepared by methods involving carbon–carbon bond formation and they can therefore be prepared by the same methods (e.g. Claisen condensation) which are commonly used for the preparation of carbocyclic compounds (scheme 11.3).

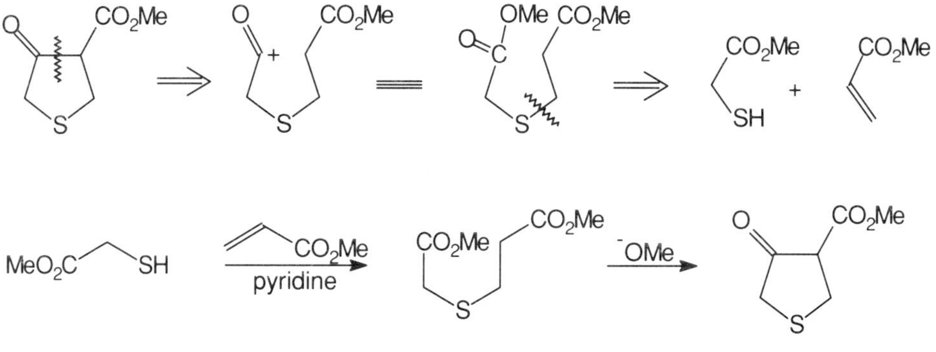

Scheme 11.3

Furthermore five-membered heterocycles can also be obtained using 1,3-dipolar cycloaddition reactions (eqns 11.18 and 11.19).

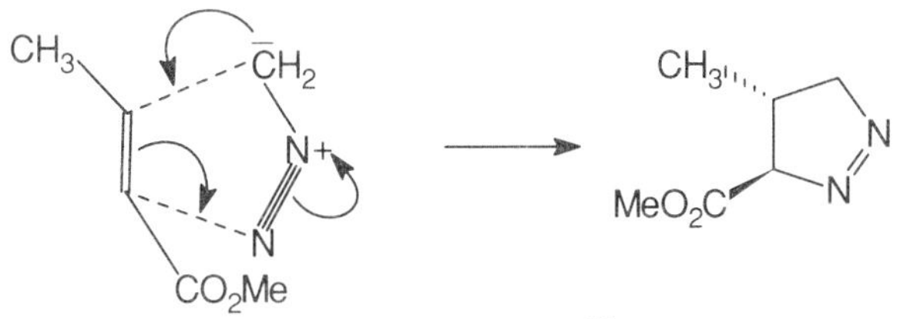

$$(11.18)$$

diazomethane

Ph—N̄—N̈⁺≡N + PhC≡CH $\longrightarrow$

phenyl azide

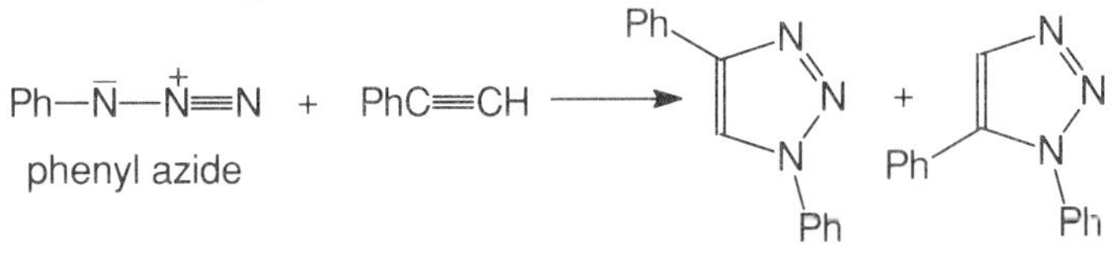

$$(11.19)$$

Six-membered rings

Saturated six-membered heterocycles can in principle be readily prepared by cyclisation of 1,5-bifunctional compounds (eqn 11.20).

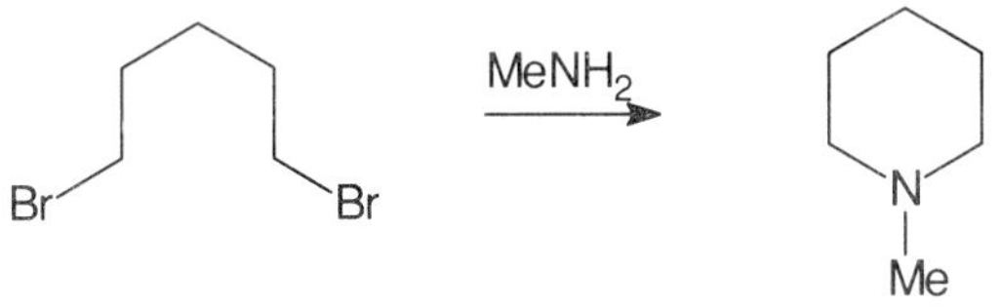

$$(11.20)$$

As with five-membered heterocyclic compounds they can also be prepared by numerous addition/condensation strategies and by methods involving carbon–carbon bond formation (e.g. eqns 11.21 and 11.22).

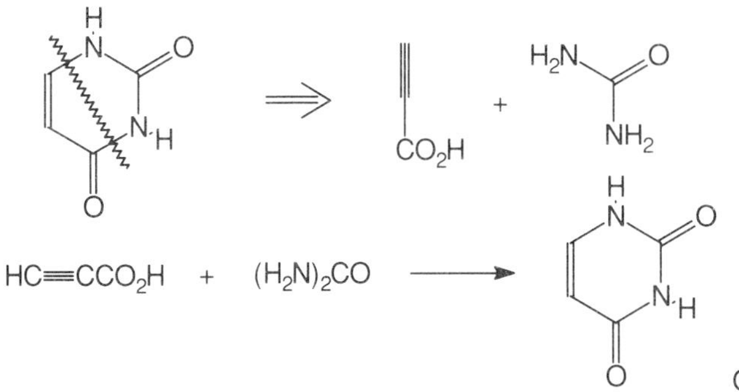

$$(11.21)$$

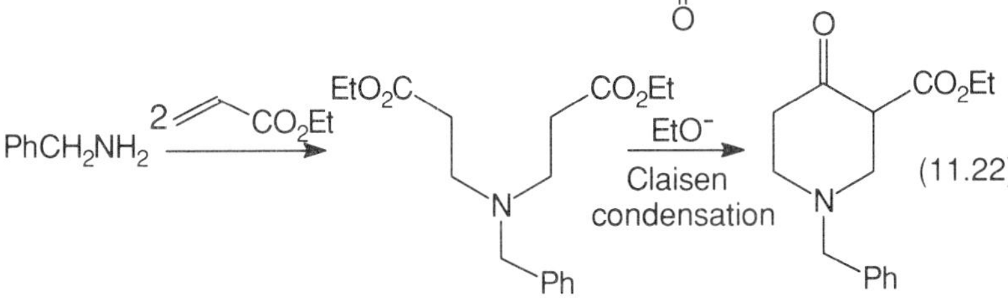

$$(11.22)$$

Furthermore, six-membered heterocyclic compounds can be prepared using hetero-Diels–Alder reactions (e.g. eqn 11.23). The catalysis of such reactions by transition metal complexes and the stereochemical control thereby achieved have been elegantly employed by Danishefsky and co-workers in the synthesis of carbohydrate precursors (eqn 11.24).

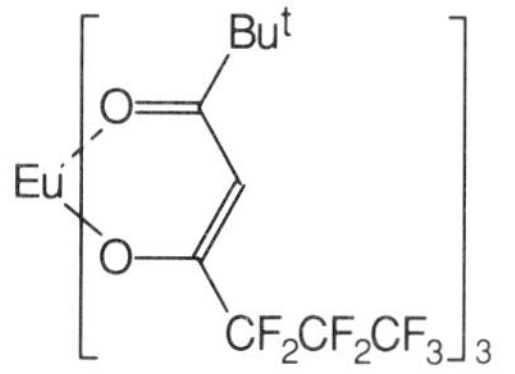

Eu(fod)$_3$ is a mild Lewis acid which is also used as an NMR shift reagent.

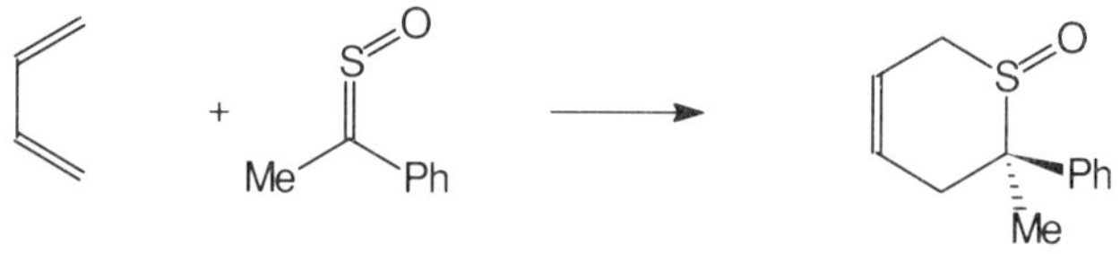

(11.23)

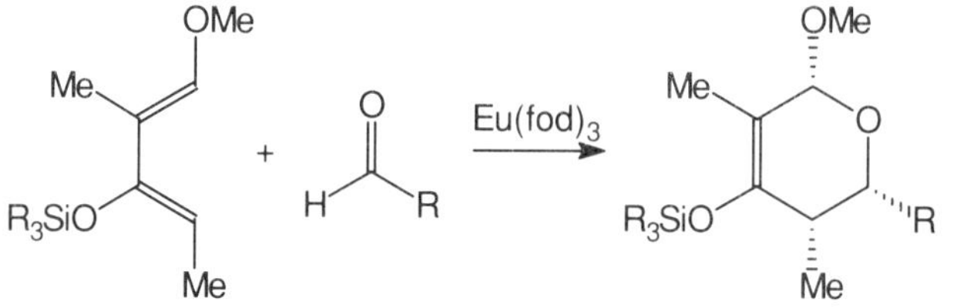

(11.24)

11.3 Polymerisation of bifunctional compounds

Addition polymers

When isoprene is treated with conventional free radical initiators it readily forms a polymer containing mainly *trans* (E) double bonds. Polymerisation using organometallic catalysts on the other hand gives a product with *cis* (Z) double bonds, as found in natural rubber (scheme 11.4).

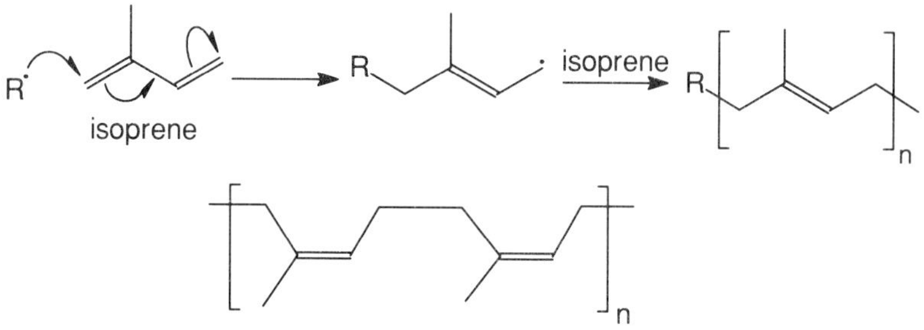

Scheme 11.4

Rubber is readily attacked by ozone and is subject to oxidative degradation when exposed to oxygen or air for long periods. However the useful properties of rubber can be enhanced and preserved if it is subjected to vulcanisation, in which a small percentage of sulphur is incorporated into the crude rubber. Carbon–sulphur bonds are formed and the improved properties are due to cross-linking of the polymer chains and removal of some of the carbon–carbon double bonds.

Polyesters and polyamides

These are formed by reacting dicarboxylic acid derivatives with diols or diamines respectively, or by self condensation of hydroxy- or amino acid derivatives (scheme 11.5).

$$MeO_2C-\underset{}{\bigcirc}-CO_2Me \ + \ HO\diagdown\diagup OH \ \longrightarrow \ \Big[CO-\underset{}{\bigcirc}-CO_2(CH_2)_2O\Big]_n$$
polyester

$$HO_2C(CH_2)_4CO_2H \ + \ H_2N(CH_2)_6NH_2 \ \longrightarrow \ \Big[CO(CH_2)_4CONH(CH_2)_6NH\Big]_n$$
polyamide
(nylon-6,6)

$$HO(CH_2)_{10}CO_2H \ \longrightarrow \ \Big[O(CH_2)_{10}CO\Big]_n$$
polyester

$$\longrightarrow \ \Big[CO(CH_2)_5NH\Big]_n$$
polyamide
(nylon-6)

Scheme 11.5

Polyurethanes and polyureas

Isocyanates react with water to give carbamic acids which spontaneously decarboxylate to give amines (eqn 11.25). They also react with alcohols to give carbamates (urethanes) and with amines to give ureas (eqns 11.26 and 11.27).

$$R-N=C=O \ + \ H_2O \ \longrightarrow \ RNHCO_2H \ \longrightarrow \ RNH_2 \ + \ CO_2 \quad (11.25)$$

$$R-N=C=O \ + \ R'OH \ \longrightarrow \ RNHCO_2R' \qquad\qquad (11.26)$$
a carbamate (urethane)

$$R-N=C=O \ + \ R'NH_2 \ \longrightarrow \ RNHCONHR' \qquad\qquad (11.27)$$
a urea

Polyurethane polymers are formed by reacting a diol with a diisocyanate (eqn 11.28). They are amorphous but have the property of forming stable flexible foams when small gas bubbles are entrapped during polymerisation. These foams are more durable than foam rubber and are widely used in upholstery and padding. Gas bubbles to provide the foam are formed by adding a small amount of water to the polymerisation mixture. Water reacts with the isocyanate group to form a carbamic acid which liberates carbon dioxide.

$$\underset{\text{toluene}\atop\text{diisocyanate}}{\overset{CH_3}{\underset{NCO}{\bigcirc}NCO}} \ + \ HO\Big[(CH_2)_4O\Big]_nH \ \longrightarrow \qquad (11.28)$$

$$\Big[NH-\underset{}{\overset{CH_3}{\bigcirc}}-NHCO_2\Big[(CH_2)_4O\Big]_nCO\Big]_m$$
polyurethane

Epoxyresins and polycarbonates

Epoxyresins are used as high strength glues (eqn 11.29). Polycarbonates have high impact strength and are used as a substitute for glass in break-resistant windows in banks and kiosks (eqn 11.30).

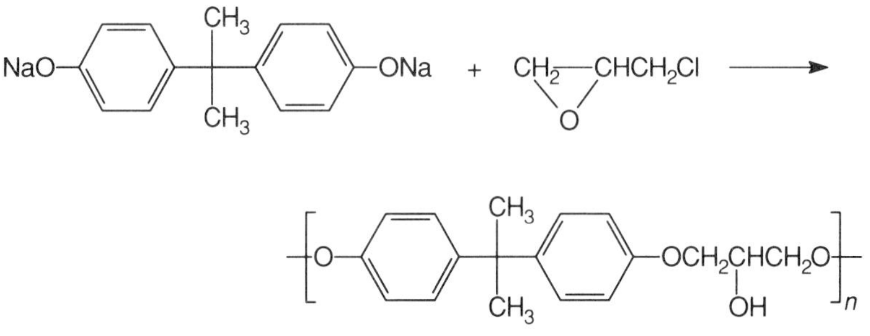

(11.29)

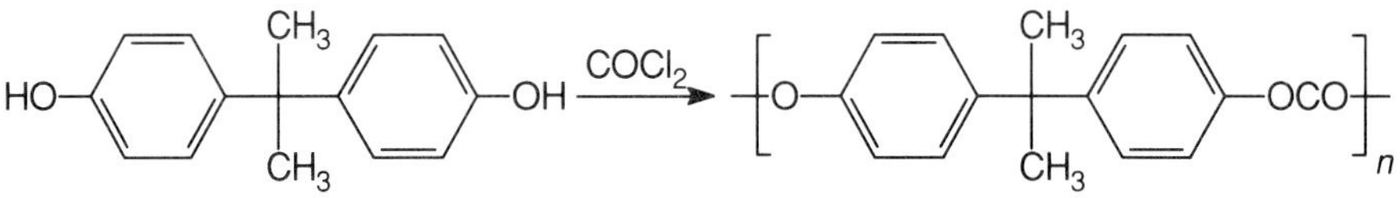

(11.30)

Problems

1. Suggest possible mechanisms for each of the following reactions.

(a)

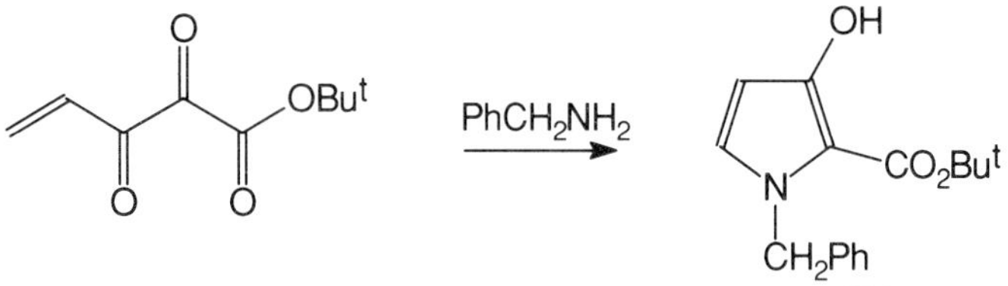

$$PhCH_2NH_2$$

(b) 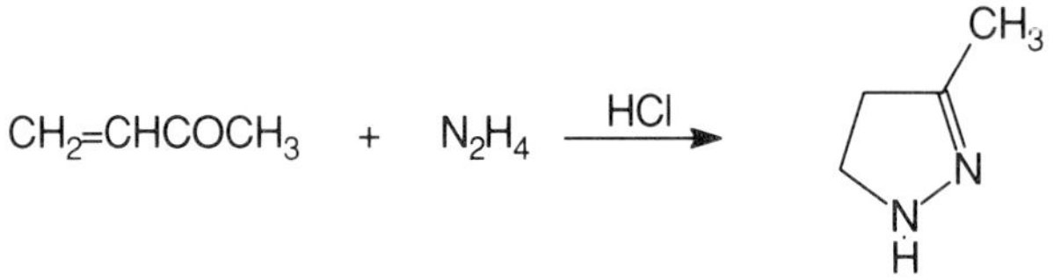 $CH_2{=}CHCOCH_3$ + N_2H_4 $\xrightarrow{\text{HCl}}$

Answers to problems

Chapter 2

1.

2.

3.

4.

Chapter 3

1.

2.

Chapter 4

1.

A   B

2. $CH_3CH_2COCH_2CH_3 \longrightarrow (C_2H_5)_2C-C(C_2H_5)_2 \longrightarrow (C_2H_5)_3C-CC_2H_5$

C   D

3.

E   F

4.

G   H

Chapter 5

1.

2. Ph

3. HO

Chapter 6

1(a).

$\xrightarrow{NaOH}$

1(b).

$\xrightarrow{NaOH}$

2(a).

$\xrightarrow{HBr}$ $\xrightarrow[NaOEt]{CH_3COCH_2CO_2Et}$

2(b).

$\xrightarrow[BrCH_2CO_2Et]{NaOEt}$ 1. $^-OH/H_2O$
2. H_3O^+/Δ
3. $EtOH/H^+$

Chapter 7

1.

2(a).

2(b).

etc.

2(c).

2(d).

Chapter 10

1(a). Ph

1(b).

2.

Chapter 11

1(a)

1(b)

Further reading

General/mechanisms
J. March (1985). *Advanced organic chemistrty* (3rd edn.), Wiley, New York.
R. T. Morrison and R. N. Boyd (1987). *Organic chemistry* (5th edn.),
    Allyn & Bacon, Boston.

General/synthesis
W. Carruthers (1986). *Some modern methods of organic synthesis* (3rd edn.),
    Cambridge University Press.
R. K. Mackie, D. M. Smith and R. A. Aitken (1982). *Guidebook to organic synthesis*
    (2nd edn.), Longman.
S. G. Warren (1982). *Organic synthesis : the disconnection approach*, Wiley,
    Chichester.

Rearrangements
L. M. Harwood (1992). *Polar rearrangements*, Oxford University Press.

Asymmetric synthesis
R. A. Aitken and S. N. Kilenyi (1992). *Asymmetric synthesis*, Chapman & Hall,
    London.
A. Koskinen (1993). *Asymmetric synthesis of natural products,* Wiley, Chichester.

Organoboron and organosilicon compounds
S. E. Thomas (1991). *Organic synthesis : the role of boron and silicon*,
    Oxford University Press.

Birch reduction/aromatic compounds
M. Sainsbury (1992). *Aromatic chemistry*, Oxford University Press.

Conjugate addition reactions
P. Perlmutter (1992). *Conjugate addition reactions in organic synthesis*, Pergamon
    Press, Oxford.

Protecting groups
T. W. Greene and P. G. M. Wuts (1991). *Protective groups in organic synthesis*
    (2nd edn.), Wiley, New York.
J. F. W. McOmie (1973). *Protective groups in organic chemistry*, Plenum Press.

Amino acid and peptide synthesis
J. H. Jones (1992). *Amino acid and peptide synthesis*, Oxford University Press.

Index